AF335293

Singularities of Caustics and Wave Fronts

Mathematics and Its Applications (*Soviet Series*)

Managing Editor:

M. HAZEWINKEL
Centre for Mathematics and Computer Science, Amsterdam, The Netherlands

Editorial Board:

A. A. KIRILLOV, *MGU, Moscow, U.S.S.R.*
Yu. I. MANIN, *Steklov Institute of Mathematics, Moscow, U.S.S.R.*
N. N. MOISEEV, *Computing Centre, Academy of Sciences, Moscow, U.S.S.R.*
S. P. NOVIKOV, *Landau Institute of Theoretical Physics, Moscow, U.S.S.R.*
M. C. POLYVANOV, *Steklov Institute of Mathematics, Moscow, U.S.S.R.*
Yu. A. ROZANOV, *Steklov Institute of Mathematics, Moscow, U.S.S.R.*

Volume 62

Singularities of Caustics and Wave Fronts

by

V. I. Arnold
Steklov Institute, Moscow, U.S.S.R.

KLUWER ACADEMIC PUBLISHERS
DORDRECHT / BOSTON / LONDON

Library of Congress Cataloging in Publication Data

```
Arnol'd, V. I. (Vladimir Igorevich), 1937-
   Singularities of caustics and wave fronts / by V.I. Arnold.
      p.    cm. -- (Mathematics and its applications. Soviet series ;
   v. 62)
   Translated from the Russian.
   Includes bibliographical references and index.
   ISBN 0-7923-1038-1 (alk. paper)
   1. Geometry, Differential.  2. Singularities (Mathematics)
I. Title.  II. Series: Mathematics and its applications (Kluwer
Academic Publishers).  Soviet series ; 62.
QA649.A75  1990
516.3'6--dc20                                        90-19243
```

ISBN 0-7923-1038-1

Published by Kluwer Academic Publishers,
P.O. Box 17, 3300 AA Dordrecht, The Netherlands.

Kluwer Academic Publishers incorporates
the publishing programmes of
D. Reidel, Martinus Nijhoff, Dr W. Junk and MTP Press.

Sold and distributed in the U.S.A. and Canada
by Kluwer Academic Publishers,
101 Philip Drive, Norwell, MA 02061, U.S.A.

In all other countries, sold and distributed
by Kluwer Academic Publishers Group,
P.O. Box 322, 3300 AH Dordrecht, The Netherlands.

Printed on acid-free paper

This book has been typeset with LATEX.

'Et moi, ..., si j'avait su comment en revenir,
je n'y serais point allé.'

Jules Verne

The series is divergent; therefore we may be
able to do something with it.

O. Heaviside

One service mathematics has rendered the
human race. It has put common sense back
where it belongs, on the topmost shelf next
to the dusty canister labelled 'discarded non-
sense'.

Eric T. Bell

Mathematics is a tool for thought. A highly necessary tool in a world where both feedback and non-linearities abound. Similarly, all kinds of parts of mathematics serve as tools for other parts and for other sciences.

Applying a simple rewriting rule to the quote on the right above one finds such statements as: 'One service topology has rendered mathematical physics ...'; 'One service logic has rendered computer science ...'; 'One service category theory has rendered mathematics ...'. All arguably true. And all statements obtainable this way form part of the raison d'être of this series.

This series, *Mathematics and Its Applications*, started in 1977. Now that over one hundred volumes have appeared it seems opportune to reexamine its scope. At the time I wrote

"Growing specialization and diversification have brought a host of monographs and textbooks on increasingly specialized topics. However, the 'tree' of knowledge of mathematics and related fields does not grow only by putting forth new branches. It also happens, quite often in fact, that branches which were thought to be completely disparate are suddenly seen to be related. Further, the kind and level of sophistication of mathematics applied in various sciences has changed drastically in recent years: measure theory is used (non-trivially) in regional and theoretical economics; algebraic geometry interacts with physics; the Minkowsky lemma, coding theory and the structure of water meet one another in packing and covering theory; quantum fields, crystal defects and mathematical programming profit from homotopy theory; Lie algebras are relevant to filtering; and prediction and electrical engineering can use Stein spaces. And in addition to this there are such new emerging subdisciplines as 'experimental mathematics', 'CFD', 'completely integrable systems', 'chaos, synergetics and large-scale order', which are almost impossible to fit into the existing classification schemes. They draw upon widely different sections of mathematics."

By and large, all this still applies today. It is still true that at first sight mathematics seems rather fragmented and that to find, see, and exploit the deeper underlying interrelations more effort is needed and so are books that can help mathematicians and scientists do so. Accordingly MIA will continue to try to make such books available.

If anything, the description I gave in 1977 is now an understatement. To the examples of interaction areas one should add string theory where Riemann surfaces, algebraic geometry, modular functions, knots, quantum field theory, Kac-Moody algebras, monstrous moonshine (and more) all come together. And to the examples of things which can be usefully applied let me add the topic 'finite geometry'; a combination of words which sounds like it might not even exist, let alone be applicable. And yet it is being applied: to statistics via designs, to radar/sonar detection arrays (via finite projective planes), and to bus connections of VLSI chips (via difference sets). There seems to be no part of (so-called pure) mathematics that is not in immediate danger of being applied. And, accordingly, the applied mathematician needs to be aware of much more. Besides analysis and numerics, the traditional workhorses, he may need all kinds of combinatorics, algebra, probability, and so on.

In addition, the applied scientist needs to cope increasingly with the nonlinear world and the extra mathematical sophistication that this requires. For that is where the rewards are. Linear models are honest and a bit sad and depressing: proportional efforts and results. It is in the nonlinear world that infinitesimal inputs may result in macroscopic outputs (or vice versa). To appreciate what I am hinting at: if electronics were linear we would have no fun with transistors and computers; we would have no TV; in fact you would not be reading these lines.

There is also no safety in ignoring such outlandish things as nonstandard analysis, superspace and anticommuting integration, p-adic and ultrametric space. All three have applications in both electrical engineering and physics. Once, complex numbers were equally outlandish, but they frequently proved the shortest path between 'real' results. Similarly, the first two topics named have already provided a number of 'wormhole' paths. There is no telling where all this is leading - fortunately.

Thus the original scope of the series, which for various (sound) reasons now comprises five subseries: white (Japan), yellow (China), red (USSR), blue (Eastern Europe), and green (everything else), still applies. It has been enlarged a bit to include books treating of the tools from one subdiscipline which are used in others. Thus the series still aims at books dealing with:

- a central concept which plays an important role in several different mathematical and/or scientific specialization areas;
- new applications of the results and ideas from one area of scientific endeavour into another;
- influences which the results, problems and concepts of one field of enquiry have, and have had, on the development of another.

This book is about caustics - and about a large part of everything else in mathematics. Let me start by quoting from the article on caustics in the (Russian) Encyclopaedia of Mathematics: 'A caustic is the envelope of rays reflected or refracted by a given curve'. The editorial comments to the annotated expanded and revised English language edition (Kluwer Academic Publishers) add the following. 'In terms of purely geometric optics these [caustics] are curves of light of infinite brightness consisting of points through which infinitely many reflected or refracted light rays pass. In reality they can often be observed as a pattern of pieces of very bright curves; e.g. on a sunny day at the seashore on the bottom beneath a bit of wavy water: or at the bottom of a cup of tea into which light is shining.

That sounds like geometric optics and, quickly, singularities and bifurcation theory but still a far cry from, say, cobordism, characteristic classes, Dynkin diagrams and Weyl groups.

However, in 1972, Arnold himself, in a seminal article, discovered that the A,D,E Dynkin diagrams appear naturally in the discussion of (certain types of) singularities. He also observed that the A,D,E diagrams occur in many different parts of mathematics and posed (in 1974) the problem of finding the common origin of all these $A-D-E$ classification results. As far as singularities are concerned the B,C,F,H diagrams turned up in the first half of the eighties and by now caustics, wave fronts, Legendre transforms, singularities of ray systems; and groups of reflections and Weyl groups are firmly linked: although not a few mysteries remain, and though Arnolds problem as to other $A-D-E$ etc. classification results (the list of these has meanwhile also grown) is far from solved.

Arnold and his students have been the main initiators and developers in all this. Arnold has also found another employ for the word 'perestroika' in this context and it is fair to say that he has caused and overseen a restructuring of singularity theory. This book is his own account of all this in his well known, classic, lucid expository syle.

<table>
<tr><td>

The shortest path between two truths in the real domain passes through the complex domain.

J. Hadamard

</td><td>

Never lend books, for no one ever returns them; the only books I have in my library are books that other folk have lent me.

Anatole France

</td></tr>
<tr><td>

La physique ne nous donne pas seulement l'occasion de résoudre des problèmes ... elle nous fait pressentir la solution.

H. Poincaré

</td><td>

The function of an expert is not to be more right than other people, but to be wrong for more sophisticated reasons.

David Butler

</td></tr>
</table>

Bussum, October 1990   Michiel Hazewinkel

Contents

Introduction

Let us start with an example: consider the distance from a point in the euclidean plane to a given curve; for instance, from an interior point of an elliptical domain to the boundary ellipse (Fig. 1). The corresponding rays (the extremals of this variational problem) are the normal lines to the ellipse. The minimal value of the functional (the distance) satisfies, as a function of the initial point, the Hamilton—Jacobi equation $(\nabla u)^2 = 1$ (at the points where it is smooth). However, this function has singularities (on the segment joining the focal points of the ellipse). The system of rays also has singularities. They lie on the astroid, which is the envelope of the system of normals of the ellipse. The envelope of a system of extremals is called the *caustic* of this system. The caustic of our system has four cusps. These singularities are stable: any curve sufficiently close to an ellipse has a caustic sufficiently close to an astroid and having four cusps.

The level lines of a solution of a Hamilton—Jacobi equation are called fronts. In our example the fronts are the equidistant curves of the ellipse. The equidistant curves close to the ellipse are smooth curves. However, the equidistants which are not so close to the ellipse have singularities. In Fig. 2 four *cuspidal singularities* on an equidistant of an ellipse are depicted. These cusps are stable: curves equidistant to any curve which is sufficiently close to the ellipse have similar singularities.

Of course, a level line of the shortest distance to the ellipse is only a part of an equidistant curve. However, most of the properties of singularities of ray systems, caustics and fronts are more transparent if we consider next to the minima also the other extremal points of the functionals. In our example we start with the study of the equidistant curves and then discern the parts we need.

For instance, consider the distance to the curve as a function of the initial point.

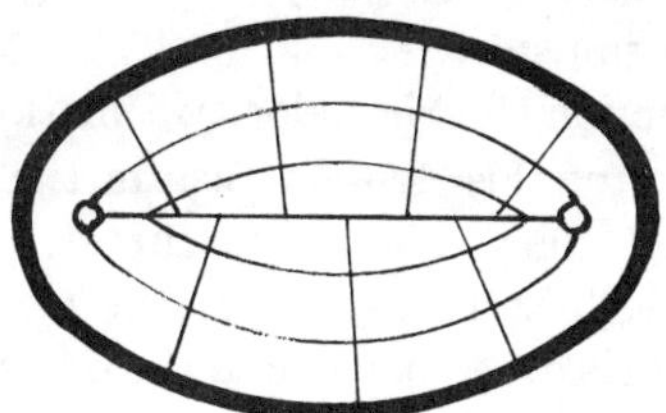

Figure 1: Rays and fronts of a perturbation, propagating inside an ellipse

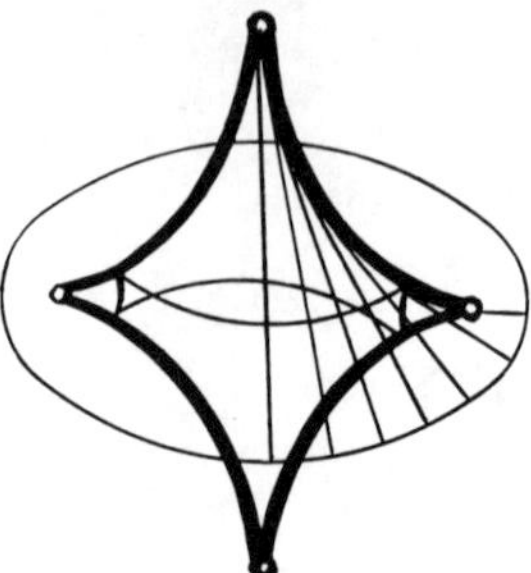

Figure 2: Caustics as envelopes of ray systems and as the set of singular points of the equidistant curves

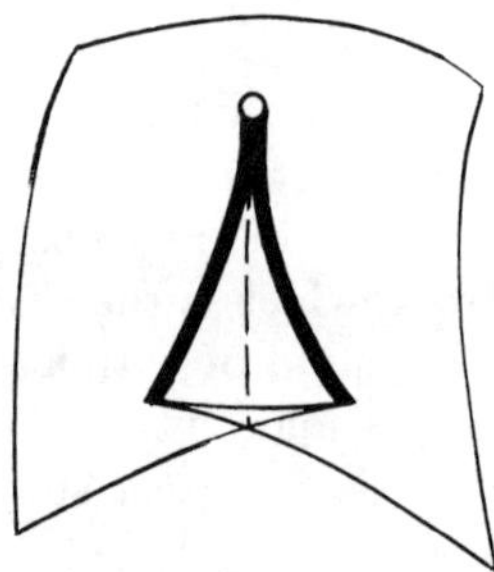

Figure 3: The swallowtail surface in the space of polynomials

It is convenient to consider it as a multivalued function, taking into account not the minimal distance alone, but also the lengths of the other normals to this curve.

The graph of the multivalued distance function (from a point inside an elliptical domain to the boundary ellipse) has a remarkable singularity at a focal point of the ellipse. Its graph is locally diffeomorphic to the *swallowtail surface* (Fig. 3). The swallowtail surface is the surface in the 3-space of polynomials

$$\mathbf{R}^3 = \{x^4 + ax^2 + bx + c\}$$

formed by the polynomials having a (real) multiple root. This singularity of the graph of the distance function is stable: by slightly perturbing the ellipse, the singularity will remain (up to a diffeomorphism).

The graph of the time function (for the velocity-1 motion from a given initial point of a domain to the boundary of this domain) lies in the space–time and intersects the isochrones of space–time along equidistant curves. Hence the perestroika[1] of the equidistant curves of the ellipse (occurring at the moment when the equidistant passes through the focal point) is diffeomorphic to the perestroika of the sections of the

---

[1] in the past the Russian mathematical term 'perestroika' was translated into English, as 'metamorphosis' or 'surgery' (like in 'Morse surgery'), but now the word has become international, and there is no need to translate it.

Figure 4: Perestroika of the sections of the swallowtail surface

swallowtail surface by the level surfaces of the time function in the three-dimensional space–time.

A generic function $f(a, b, c)$ in this 3-space containing a swallowtail surface is reducible to the normal form $f = \pm a + \text{const}$ by a swallowtail preserving local diffeomorphism (in a neighborhood of the origin $a = b = c = 0$; the genericity condition is: $\partial f/\partial a|_0 \neq 0$, see [1]). Hence the perestroikas of the equidistant curves are the same as the perestroikas of the planar sections of the swallowtail surface $a = \text{const}$ (Fig. 4). The study of these sections is much easier than the initial problem of studying generic perestroikas of fronts. In order to study the minimum points, it suffices to delete a part of the swallowtail surface, namely the pyramid whose edges are the cuspidal edges and the line of selfintersection of the swallowtail surface.

The principles of the application of singularity theory to the study of ray systems, wave fronts, caustics, and their perestroikas in general variational problems are similar to what we have seen in the previous example.

1. There exists a list of standard singularities (like the cusps and swallowtails in the example). These singularities are (rather mysteriously) related to the geometry of reflection groups. One can study them using appropriate algebraic means (Lie groups, invariant theory, root systems, Dynkin diagrams, etc.).

2. One proves that the standard singularities are stable, and that they are the only singularities that one encounters generically. For instance, the caustic of a circle consists of one point—its center. This is not a standard singularity of caustics. But if we perturb the circle a little, this degenerate singularity of the caustic at the center is transformed into a small caustic curve whose only singularities are generic cusps and selfintersection points.

3. In this way the study of singularities and perestroikas of such objects as wave fronts, caustics, ray systems, etc., related to a generic variational problem is reduced to the study of standard singularities for the corresponding objects. For instance, the study of perestroikas of moving wave fronts is reduced to the study of sections of (generalized) swallowtails by hypersurfaces.

This book is an account of the main results obtained in the realisation of the above program since 1972, when the relation between singularities of ray systems, their caustics, wave fronts, Legendre transforms, and reflection and Weyl groups was discovered [2]. The groups A, D, E (having only simple lines in their Dynkin diagrams) appeared first. Subsequently, in 1978 the relation between the groups with

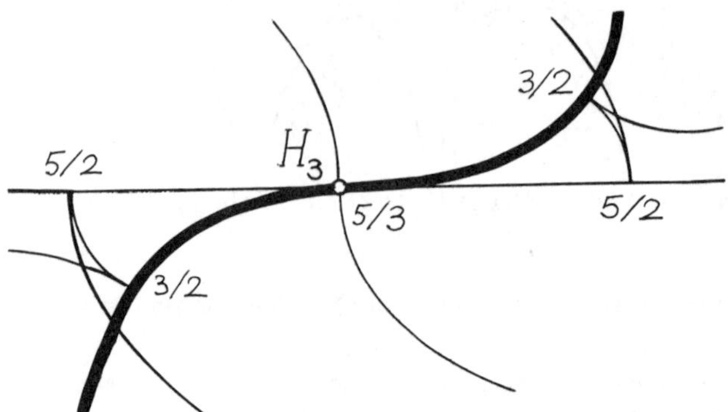

Figure 5: The system of involutes at an inflection point of a curve

double lines (B, C, F) and boundary singularities (for instance, the singularities of the distance function to a manifold with boundary) appeared [3].

The search for other reflection groups (H_2, H_3, H_4; $I_2(p)$) started immediately. During the fall of 1982 the joint efforts of O.V. Ljashenko, A.B. Givental, O.P. Shcherbak, and the author led to the discovery of the icosahedron symmetry group H_3; it controls the singularities of the ray system and the fronts in the variational problem of fastest bypassing of a plane obstacle bounded by a generic curve with an inflection point ([4], [5]). In other words, this group governs the family of involutes of a plane curve having an inflection point (like a cubic parabola), Fig. 5.

O.P. Shcherbak has found in 1984 the most complicated 'hypericosahedron' H_4, related to the singularity in the obstacle problem in 3-space ([6], [7]).

The techniques on which the study of singularities of ray systems, caustics, wave fronts, and other objects related to geometrical optics and the calculus of variations rests, come from symplectic and contact geometry. Symplectic analysis of Shcherbak's work has led A.B. Givental [8] to the discovery of the classification problem in the geometry of wave front systems whose solutions are in one-to-one correspondence with the finite Coxeter groups, generated by euclidean reflections.

While most of the results discussed in this book were obtained during the last two decades, the subject itself is very classical and goes back to Archimedes, Huygens, Barrow,...; it is rather strange that the classical scientists have missed these results. For instance, the local classification of projections of generic smooth surfaces from points of ordinary 3-space was discovered only in 1981 (O.A. Platanova [9], O.P. Shcherbak). There exists a finite list of inequivalent germs of such projections; it contains 14 projections. This is thus the number of different shapes of a neighborhood of a point on a generic surface, seen from the different points of space.

One possible explanation of the late discovery of the fundamental classification theorems is that the proofs are difficult and depend on unexpected relations between many branches of mathematics. In order to understand the unexpected cancellation of many terms in dull and long computations, the strange similarity of bifurcation diagrams in apparently unrelated problems and the mysterious appearance of the regular polyhedra in problems of applied mathematics, one has to replace the straightforward computations in differential geometry by the simple and general approach of symplectic and contact geometry.

The general principle here is to lift the geometrical objects from the 'configuration' space V to the 'phase' space T^*V, in which the singularities either disappear or simplify (in the theory of partial differential equations and in quantum theory this approach is called the 'microlocal' study). This lifting transforms simple facts of differential geometry to general theorems of symplectic and contact geometry, whose range of applicability is much wider (for instance, one may use the differential-geometric intuition on surfaces in euclidean space in order to obtain results on general variational problems with one-sided constraint).

Hence I will start with a short account of the well-known elementary facts in symplectic and contact geometry. However, I do hope that the reader will be able to understand at least the statements of the results he needs, even if he would not have mastered some of the foundations preceding them.

Chapter 1

Symplectic geometry

Symplectic geometry is the geometry of phase space. This chapter contains some standard definitions and facts from elementary symplectic geometry, together with some less familiar examples (for instance, the symplectic structures of spaces of polynomials, and the theory of normal forms of submanifolds of a symplectic manifold).

1.1 Symplectic manifolds

A *symplectic structure* on a manifold is a closed nondegenerate 2-form (also called a *symplectic form*).

Example 1 (*area*). The area element $dp \wedge dq$ defines on the (p, q)-plane a symplectic structure.

Example 2 (*products*). The direct product of two symplectic manifolds inherits a natural symplectic structure from the factors.

Example 3 (*normal form*). $\mathbf{R}^{2n}$ has, for instance, the symplectic structure of a product of n planes, $\sum dp_i \wedge dq_i$.

Theorem (Darboux). *All symplectic manifolds of a given fixed dimension are locally symplectically diffeomorphic.*

In other words, any closed nondegenerate 2-form can be represented in the local normal form of example 3, using appropriate local '*Darboux coordinates*'.

Remark. The dimension of a symplectic manifold is even, since a skewsymmetric form in an odd-dimensional space is degenerate (any rotation of an odd-dimensional space has an axis).

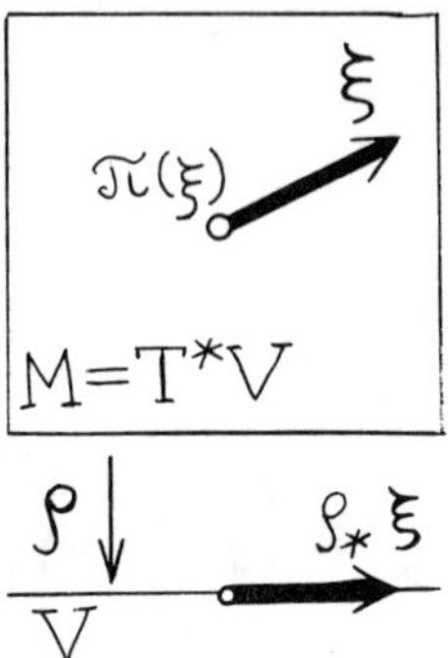

Figure 6: Definition of the action form

Two vectors based at the same point of a symplectic manifold are called *skeworthogonal* if the value of the symplectic form on these vectors is zero. For instance, in Darboux coordinates all p axes are skeworthogonal.

Example 4 (*phase space*). The *phase space* (the total space of the cotangent bundle) $M = T^*V$ of any smooth manifold V is equipped with a natural symplectic structure, expressed as $\omega = \sum dp_i \wedge dq_i$ in the ordinary coordinates of the phase space.

It is important to note that both the 'Poincaré integral invariant' $\omega = d\alpha$ and the 'action form' $\alpha = \sum p_i \, dq_i$ do *not* depend on the special choice of local coordinates q_i on V. In other words, the diffeomorphisms of V map α to α and ω to ω.

To prove this it suffices to note that the action form α has a simple coordinate-free description. Namely,

$$\alpha(\xi) = \pi(\xi)\rho_*(\xi) \qquad (\xi \in TM),$$

where $\pi : TM \to M$ is the tangent bundle of the phase space and

$$\rho : T^*V \to V$$

is the cotangent bundle of the configuration space (Fig. 6).

Let us consider a hypersurface in a symplectic manifold.

Definition. The *characteristic direction* at a point of a hypersurface is the skeworthogonal complement to the tangent space (at this point) (Fig. 7).

The skeworthogonal complement to a hyperplane lies in the hyperplane. Hence a characteristic direction is tangent to the hypersurface. Thus, the hypersurface is equipped with a field of characteristic directions.

The integral curves of this field are called the *characteristics* of the hypersurface.

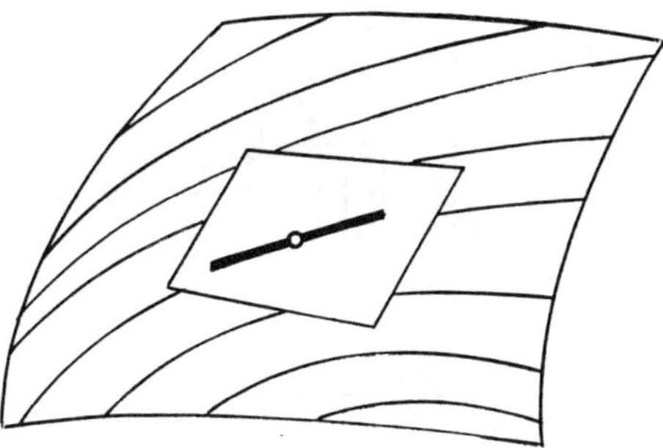

Figure 7: A characteristic direction and characteristic on a hypersurface in a symplectic space

Example 5 (*characteristics*). The characteristics of the hypersurface $H(p,q) = 0$ (where $dH \neq 0$ and (p,q) are Darboux coordinates) are the trajectories of the Hamilton equation

$$\dot{p} = -\frac{\partial H}{\partial q}, \qquad \dot{q} = \frac{\partial H}{\partial p}.$$

They are also the extremals of the variational principle $\delta \int \alpha = 0$, $H = 0$, where $\alpha = p\,dq$.

Remark. In mathematics the word 'characteristic(al)' invariably means 'intrinsically related to' or 'independent of arbitrary choices'. Thus, the characteristic equation of an operator is independent of the coordinate system, a characteristic subgroup is invariant under automorphisms (of the whole group), and characteristic cohomology classes are invariant under diffeomorphisms.

The characteristic direction (characteristic) of a hypersurface in a symplectic space at a point may be defined as the only direction that is invariant under the symplectomorphisms (symplectic diffeomorphisms) preserving the hypersurface and the point. To prove this it suffices to consider any particular hypersurface—say the hyperplane defined in Darboux coordinates by the equation $p_1 = 0$—since all hypersurfaces in a symplectic manifold are locally symplectomorphic (as we will soon see).

Consider the space of characteristics of a hypersurface in a symplectic manifold. Globally this space is not well-defined, but locally (in a neighborhood of any point of the hypersurface) the characteristic directions form a manifold. The dimension of this manifold is two less than the dimension of the initial symplectic space, and is thus even.

Example 6 (*manifold of extremals*). The manifold of characteristics (characteristic manifold) of a hypersurface in a symplectic manifold inherits a symplectic structure.

Namely, the symplectic structure of the space of characteristics is defined by the

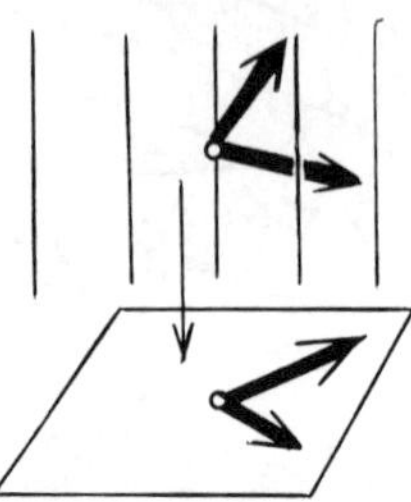

Figure 8: The symplectic structure of a characteristic space

following condition: the value of the initial symplectic form on two arbitrary vectors based at the same point and tangent to the hypersurface is equal to the value of the inherited structure on the projections of these vectors on the space of characteristics (Fig. 8).

Remark. The fact that this condition determines a symplectic structure is a useful exercise. It is also useful to note that the symplectic characteristic manifold is obtained from the initial manifold in two steps.

In ordinary geometry there are two ways to decrease the dimension of a manifold—sectioning and projection. In symplectic geometry the dimensions of symplectic manifolds are even, and decrease of dimension is always achieved in two steps, one of which is sectioning while the other is projection. To obtain the characteristic manifold we start with sectioning by a codimension-1 section (a hypersurface) and then project it with 1-dimensional kernel onto the characteristic manifold.

Example 7 (*space of lines*). Consider in the phase space $\mathbf{R}^{2n}$ defined over the euclidean space $\mathbf{R}^n$ the hypersurface which defines the euclidean metrics, $p^2 = 1$. The characteristics of this hypersurface correspond to the motions of a free particle. Hence we can identify the characteristics with oriented straight lines in $\mathbf{R}^n$. Thus, the characteristic manifold of this particular hypersurface is globally well-defined: it is the *manifold of all oriented lines in* $\mathbf{R}^n$.

This manifold may be identified with the total space of the (co)tangent bundle of the sphere S^{n-1}. First we associate to each oriented line its unitary velocity vector $v \in S^{n-1}$. The point of intersection of the line with the hyperplane of $\mathbf{R}^n$ tangent to S^{n-1} at v defines the required element of T^*S^{n-1} (Fig. 9). This identification equips the characteristic manifold with a new symplectic structure (inherited from the natural symplectic structure of T^*S^{n-1} as the phase space).

These two symplectic structures of the characteristic manifold of our hypersurface (inherited from $T^*\mathbf{R}^n$ or from T^*S^{n-1}) differ only in sign.

Remark. In quite the same way does the 'manifold' of extremals of a regular 1-dimensional variational problem corresponding to a fixed energy level inherit a natural symplectic structure. (This is usually underestimated in traditional expositions of the

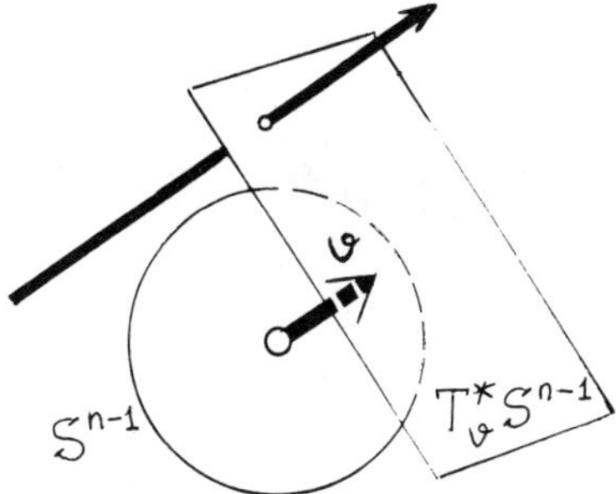

Figure 9: The manifold of oriented lines in a euclidean space

calculus of variations and optimal control theory.)

The following examples of symplectic structures in spaces of polynomials are very important, since they provide us with useful normal forms of singularities of Lagrangian and Legendre varieties.

Example 8 (*binary forms*). Consider the linear space V of real forms (homogeneous polynomials) of degree d in two variables x, y. The dimension of this space is $d+1$. If d is odd, V is even-dimensional. The group $G = \mathrm{SL}(2, \mathbf{R})$ (of linear transformations of the plane with determinant 1) acts on V. There exists a nondegenerate skewsymmetric G-invariant form on V. This form is, essentially, unique (up to a nonzero factor; this was more-or-less known to Hilbert).

Hence there exists a natural symplectic structure on the space of odd-degree binary forms (defined up to a nonzero factor).

An explicit description of this symplectic structure is given by the following formula: a binary form ϕ of degree $d + 1$ can be expressed as

$$\phi(x, 1) = q_0 e_d + \cdots + q_n e_{n+1} - p_n e_n + \cdots - (-1)^n p_0,$$

where $e_s = x^s / s!$, $(p_0, \ldots, q_n)$ are appropriate Darboux coordinates, and the signs of the terms involving p's alternate.

Example 9 (*even-degree polynomials*). The hyperplane $q_0 = \mathrm{const}$ may be identified with the space of polynomials in x of degree d and with fixed leading coefficient. The function q_0 is the Hamilton function of the translations along the axis of the values of a polynomial. Hence the symplectic manifold of characteristics of this hypersurface may be identified with the manifold of polynomials of even degree $b = d - 1 = 2n$ and with fixed leading coefficient,

$$f(x) = e_b + q_1 e_{b-1} + \cdots + q_n e_n - p_n e_{n-1} + \cdots - (-1)^n p_1,$$

where $(p_1, \ldots, q_n)$ are Darboux coordinates.

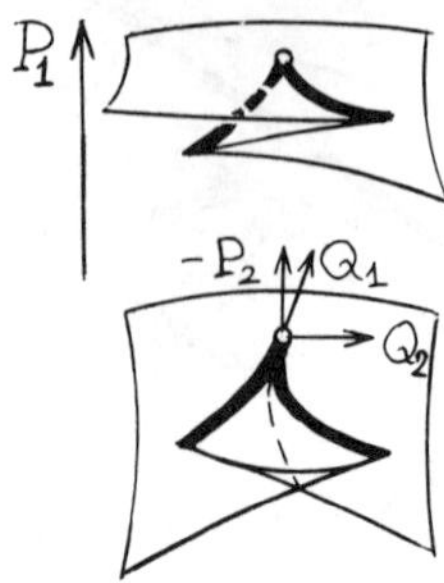

Figure 10: The unfurled swallowtail over the ordinary swallowtail

Example 10 (*odd-degree polynomials of trace 0*). Translations along the x-axis preserve the symplectic structure in example 9. They form a 1-parameter group of symplectomorphisms. The Hamilton function of this group is (as one easily sees)

$$h = p_1 + q_1 p_2 + \cdots + q_{n-1} p_n + \frac{1}{2} q_n^2.$$

(This formula was known to Hilbert [10], see [11], [12].)

Consider the characteristic manifold of the hypersurface $h = 0$. We identify this manifold with the manifold of polynomials of odd degree $a = b - 1$ and with fixed leading coefficient and zero trace (i.e. sum of the roots). Indeed, the translation reducing the trace to zero is uniquely defined, and the value of p_1 is defined by the values of the other coefficients, since $h = 0$.

Thus we have defined a symplectic structure in the space of odd-degree polynomials $g(x)$ with fixed leading coefficient and subsequent coefficient zero:

$$g(x) = e_a + Q_1 e_{a-2} + \cdots + Q_m e_m - P_m e_{m-1} + \cdots + (-1)^m P_1,$$

where $a = 2m + 1$ and $(P_1, \ldots, Q_m)$ are Darboux coordinates.

For instance, consider the space of degree-5 polynomials

$$\frac{1}{120} x^5 + \frac{Q_1}{6} x^3 + \frac{Q_2}{2} x^2 - P_2 x + P_1.$$

The polynomials of this form having a triple real root form a (singular) surface in this symplectic 4-space (Fig. 10). This surface (called the *open swallowtail* or the *unfurled swallowtail*) is a Lagrangian variety, that is, the symplectic structure vanishes on this surface.

The unfurled swallowtail was the first nontrivial example of a singular Lagrangian variety, and is important in the theory of singularities of caustics and wave fronts; it is related to the icosahedron hidden at an inflection point of the boundary of an obstacle in the euclidean plane.

1.2 Submanifolds of symplectic manifolds

The submanifolds of a euclidean or riemannian manifold are distinguished one from another both by the interior geometry and by position in the ambient space (for instance, a surface in euclidean 3-space has, besides Gaussian curvature, also mean curvature). In the symplectic case the situation is simpler: the interior geometry of a submanifold determines (at least locally) the exterior geometry.

Theorem (A.B. Givental, [13], [14], [8]). *A germ of a submanifold of a symplectic space is defined, up to a symplectomorphism, by the restriction of the symplectic structure to its tangent bundle.*

This theorem generalises Darboux's theorem (which corresponds to the particular case when the submanifold is a point). It is closely related to the following generalisation of Darboux's theorem:

Theorem (A. Weinstein, [15]). *A submanifold of a symplectic space is defined, up to a symplectomorphism of a neighborhood of it, by the restriction of the symplectic form to the bundle of tangent vectors of the ambient space at the point of the submanifold.*

Weinstein's theorem is global in the sense that it gives a conclusion on the whole submanifold. Givental's theorem is local; however, it requires only 'interior' information, whereas in order to be able to apply Weinstein's theorem we must know some exterior information (namely, the values of the symplectic form on nontangential elements).

For our purposes Givental's theorem is extremely useful, and Weinstein's theorem is rather useless.

Proof. It suffices to prove that given two arbitrary symplectic structures which coincide on a germ of a submanifold, one can be transferred to the other by a diffeomorphism preserving the germ. We will even prove more: one may choose this diffeomorphism so as to leave invariant every point of the submanifold.

We start with the linearised problem: the restrictions of the symplectic structures to the tangent plane of the submanifold at the origin of the germ coincide.

Lemma 1. *The only symplectic invariants of a subspace of a linear symplectic space (equipped with a nondegenerate skewsymmetric bilinear form) are its dimension and the rank of the restriction of the form to the subspace.*

In other terms, given two arbitrary planes of the same dimension and rank, there exists a linear symplectic transformation of the ambient space transforming the first plane to the second. This transformation may be chosen to coincide with an arbitrary linear map of the first plane to the second, provided that this map transforms the restriction to the first plane of the symplectic form into the restriction to the second

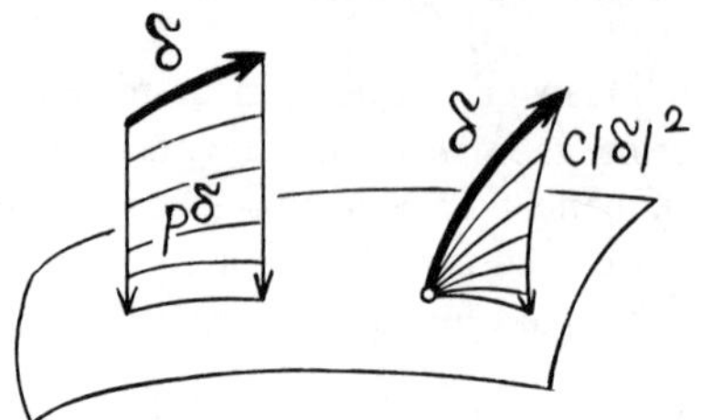

Figure 11: The relative Poincaré lemma

plane.

The very easy proof of this lemma from linear algebra is left to the reader (it is similar to the proof of the corresponding statement in ordinary euclidean geometry).

Lemma 2. *For a given smooth 1-parameter family of symplectic structures ω_t that coincide on a germ of a submanifold there exists a family of local diffeomorphisms leaving invariant the points of the submanifold and transforming ω_0 into ω_t for any t.*

The theorem follows from lemmas 1 and 2. Indeed, by lemma 1 we choose a diffeomorphism leaving invariant the points of the submanifold and transforming the second symplectic structure to the first one, at the origin. Let the two structures which coincide at the origin be ω_0 and ω_1. Then the family $\{\omega_t = \omega_0 + t(\omega_1 - \omega_0)\}$ is constant at the origin. Thus the forms ω_t, $0 \leq t \leq 1$, are nondegenerate at the origin and hence define symplectic structures in a neighborhood of it.

Lemma 2, applied to this family, provides the required equivalence.

Proof of lemma 2. We try to find the required family of diffeomorphisms as a family of transformations of time t defined by a time-dependent family of vector fields V_t. For V_t we obtain a '*homological equation*', namely

$$L_{V_t}\omega_t = \beta, \qquad \text{where } \beta = -\frac{\partial \omega_t}{\partial t}.$$

Here L_{V_t} is the Lie derivative, $L_V = di_V + i_V d$. Since ω_t is closed, the homological equation reduces to

$$di_{V_t}\omega_t = \beta.$$

Now β is closed, and hence locally $\beta = d\gamma$ (the Poincaré lemma).

We may choose a 1-form γ such that it vanishes at all points of the submanifold. This 'relative Poincaré lemma' follows from the local contractibility of a neighborhood of a submanifold to the submanifold.

Indeed, the values of this 1-form γ can be explicitly written as integrals along the orbits of the contractions; namely, $\int_\delta \gamma = \int_{p\delta} \beta$, where $p\delta$ is the trace swept by the chain δ while contracting the neighborhood to the submanifold (Fig. 11). The area sweeped by a 1-chain δ starting at a point of the submanifold is majorised by $C|\delta|^2$. Hence γ vanishes at the points of the submanifold.

Thus, the homological equation becomes

$$di_{V_t}\omega_t = d\gamma,$$

where the 1-form γ vanishes at all point of the submanifold. Since the forms ω_t are nondegenerate, the linear algebraic equations

$$i_{V_t}\omega_t = \gamma$$

define unique solutions V_t. These vector fields depend smoothly on t and vanish on the submanifold. The time-dependent differential equation $\dot{x} = V_t(x)$ defines the required difffeomorphisms (sending the initial value of x at time $t = 0$ to the value of the corresponding solution at time t). These are defined locally, but since V vanishes at the origin, the diffeomorphisms are well-defined for $0 \leq t \leq 1$ in some neighborhood of the origin.

Example 1. All symplectic submanifolds of given, fixed, dimension of a symplectic space are locally symplectomorphic (by a local symplectomorphism of the ambient space).

Example 2. All Lagrangian submanifolds of a given symplectic space are locally symplectomorphic (and hence locally symplectomorphic to any particular one, say the p-plane in the Darboux model).

Example 3. All submanifolds for which the restrictions of the symplectic structure of the ambient space have fixed, constant, rank are locally symplectomorphic. (As a local model we may choose, say, a coordinate plane $(p_1, \ldots, p_r; q_1, \ldots, q_{r+s})$ in the Darboux model; the rank of a skew form is always even, $2r$.).

Example 4. All hypersurfaces of a symplectic manifold are locally symplectomorphic.

Indeed, the skeworthogonal complement of a hyperplane is 1-dimensional and lies in the hyperplane. Thus the corank of the restriction of the symplectic structure to a hypersurface at a point equals 1. Hence the corank is constant, and 'all hypersurfaces are alike'. In the same way, 'all curves are alike'.

Example 5. Consider a generic 2-dimensional oriented surface in a symplectic manifold (of higher dimension). The restriction of the symplectic structure to the surface is a 2-form, $f\tau$, where τ is (some) area element and f is a smooth function.

The germs of the surface at points outside the line $f = 0$ are symplectic (example 1). At the points of the degeneration line $f = 0$ the restriction of the form vanishes.

Generically, a degeneration line is smooth. At a nondegenerate degeneration point ($f = 0$, $df \neq 0$) the 2-form is reducible to normal form $x\,dx \wedge dy$ by a suitable choice of local coordinates. Thus Givental's theorem implies the following.

Corollary 1. *A generic 2-surface in a symplectic 4-manifold is locally reducible by a symplectomorphism to one of the following two normal forms:*

$$p_2 = q_2 = 0 \qquad \textit{(generic points)};$$

$$p_1 - p_2^2 = q_2 = 0 \qquad \textit{(degeneration points)}.$$

To obtain the local normal forms of a generic 2-surface in a symplectic space of higher dimension, it suffices to add the equations $p_3 = q_3 = \cdots = 0$.

Example 6. Consider an even-dimensional submanifold of a symplectic space. The restriction of the symplectic structure to this submanifold is generically degenerate at some *degeneration hypersurface* of the submanifold. At a generic point of the degeneration hypersurface the restriction is reducible to the normal form

$$x_1 \, dx_1 \wedge dy_1 + dx_2 \wedge dy_2 + \cdots + dx_n \wedge dy_n$$

(this is left as an (easy) exercise).

Corollary 2. *A generic even-dimensional submanifold of a symplectic 2n-manifold is, in a neighborhood of a generic point of its degeneration hypersurface, reducible to the normal form*

$$p_1 - p_2^2 = q_2 = p_3 = q_3 = \cdots = p_k = q_k = 0.$$

Remark. These results hold both in the smooth and in the analytic (or holomorphic) case. The theory of higher degenerations is more complicated: the series leading to normal forms diverge, in general.

In the following examples the words 'functions', 'forms', 'diffeomorphisms' mean C^∞ objects (even if the initial data which is reduced to normal form is analytic).

Example 7. Consider a 4-dimensional submanifold in a symplectic manifold of dimension 6 (or higher). The degeneration points (of the restriction of the symplectic structure to the submanifold) generically form a smooth degeneration hypersurface of dimension 3. The rank of this restriction is, in general, equal to 2 at the points of this 3-surface (one needs $6 > 3$ parameters to obtain a point of corank 4; $2c^2 - c$ parameters to obtain a point of corank $2c$).

Hence the degeneration 3-surface is equipped with a smooth field of 2-planes tangent to the 4-manifold (the kernels of the restrictions). At a generic point of the degeneration 3-surface the kernel plane is transversal to this surface. The intersection line is the kernel of the restriction of the symplectic structure to the degeneration 3-surface. Thus, this 3-surface is equipped with a field of tangent directions, defined intrinsically by the symplectic structure; it is called the characteristic field.

However, at some points of the degeneration 3-surface the kernel planes may be tangent to the 3-surface. For generic 4-manifolds this happens on some line on this 3-surface. The points of this line are singular for the characteristic field. The restriction of the symplectic form to the 3-surface vanishes on this line.

The singularities of closed degenerate 2-forms on 4-manifolds on such lines were studied by J. Martinet [16], and are called *Martinet singularities*. He has proved that generically (for generic forms and at generic points of the singular lines) the 2-form is reducible by a diffeomorphism to one of two normal forms (called the *elliptic* and *hyperbolic Martinet models*). In order to describe these models we choose a volume element (a nondegenerate 3-form) on the degeneration 3-surface. The restriction of our closed 2-form to this surface is the flow of a divergence-free vector field (depending on the choice of the volume element; however, its direction field, being the characteristic direction field, is defined intrinsically).

This vector field vanishes on the line of Martinet singularities. Now we will reduce this divergence-free vector field in 3-space to normal form.

The general problem of the normal form of a divergence-free vector field in 3-space at an isolated zero of the field is almost as difficult as the problem of celestial mechanics. However, in our case the field is even more degenerate: the zeros are not isolated, but form a line. It turns out that the problem of reducing (by a diffeomorphism of the space followed by multiplication of the field by a suitable factor) a divergence-free vector field in 3-space on a singular line is much more easy than the analogous problem for generic divergence-free vector fields.

Consider the linear part of the field at the origin (chosen to be a fixed point). One of the three eigenvalues and the sum of all eigenvalues are equal to zero. Hence the other two eigenvalues are either real and of opposite sign, or are complex conjugate. At a generic point of the singular line they are not equal to zero. Hence the linear part is reducible (by a suitable choice of coordinates accompanied by multiplication of the vector field by a suitable function) to one of the two normal forms $x\,\partial_x - y\,\partial_y$ or $x\,\partial_y - y\,\partial_x$ (at the origin and at points close to the z-axis). In the first case (real eigenvalues) the singularity is called hyperbolic, in the second—elliptic.

The analysis of higher-order terms in the Taylor series now follows the standard scheme of the theory of normal forms (see, e.g., [17]). The final normal forms for the fields are:

$$x\,\partial_x - y\,\partial_y + xy\,\partial_z, \qquad x\,\partial_y - y\,\partial_x + (x^2 + y^2)\,\partial_z$$

(in the hyperbolic, respectively elliptic, case).

These formulas indicate that, e.g., in the elliptic case the integral curves of the normalised field of the kernels of the 2-form are helical lines with as axes the line of singularities. The curls of a helix close to this axis are almost circles, which lie very close to each other (the drift along the line is proportional to the radius squared), Fig. 12.

The field of kernels on the 3-surface defines the symplectic type of the germ of the initial 4-manifold. Indeed, computation shows that there is no information loss at any step of the construction: a field of kernels defines a closed 2-form on the 3-surface up to a diffeomorphism of this 3-surface, the restriction of a closed 2-form

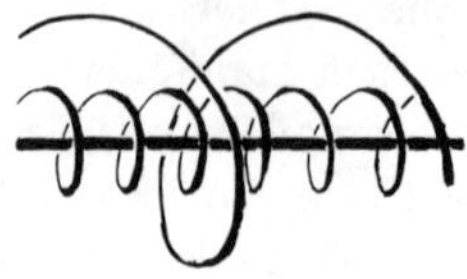

Figure 12: Typical degeneration of a symplectic structure on a 4-manifold

to a degeneration 3-surface defines the germ of a closed 2-form on the 4-space up to a diffeomorphism of the 4-space, and, finally, a germ of the closed 2-form on a 4-submanifold of a symplectic space induced from the symplectic structure defines the submanifold up to a symplectic diffeomorphism (by Givental's theorem).

The final normal forms of 4-submanifolds of a symplectic space are:

$$p_2 = p_1 p_3 \pm q_1 q_2 + \frac{q_3^3}{6}, \qquad p_3 = 0, \qquad p_4 = q_4 = \cdots = 0$$

(in Darboux coordinates).

Remark 1. Similar formulas define the codimension-3 singularities on any generic even-dimensional submanifold of a symplectic manifold (replace 4 by k in the normal forms to obtain a submanifold of dimension $2k - 4$).

The normal forms of degenerations of symplectic forms on generic submanifolds of dimension 6 (or higher) form a list that is longer than the above one. However, the above singularities are stable (do not disappear under small deformation of the submanifold; an equivalent singularity is formed at a nearby point under deformation) and simple (all nearby singularities form a finite list of equivalence classes).

Our list contains all simple stable germs of (even-dimensional) submanifolds of symplectic manifolds (the only simple odd-dimensional submanifolds are those of constant rank). For instance, the elliptic and hyperbolic parts of a singular line on a 4-submanifold are separated by parabolic points, which are not simple. Hence a generic 4-submanifold with a parabolic point is symplectically unstable (it is not symplectomorphic to slightly perturbed submanifolds). Such unstable 4-submanifolds form an open set in the space of 4-submanifolds.

Remark 2. The local classification of submanifolds of a symplectic space and that of degenerations of closed 2-forms are completely equivalent if the dimension of the space is not fixed. Indeed, any closed 2-form on an n-manifold is locally the differential of a 1-form $\alpha = f_1 \, dq_1 + \cdots + f_n \, dq_n$. This 2-form is induced from the standard form $dp \wedge dq$ in the Darboux space by the embedding $q \mapsto (q, p = f(q))$. Hence any closed 2-form on an n-dimensional manifold is locally induced from a symplectic $2n$-space. Two closed forms are locally reducible to each other by a diffeomorphism of the n-manifolds if and only if the corresponding submanifolds of the symplectic $2n$-spaces are symplectomorphic (by Givental's theorem).

Remark 3. In applications it is sometimes important to reduce to normal form certain subvarieties with specific singularities in a symplectic space. For instance, the diffraction problem in optics and the obstacle bypassing problem in optimal control theory lead to the symplectic classification of pairs of transversal hypersurfaces in a symplectic space.

The final answer is the (C^∞) normal form, found by Melrose [18]: in Darboux coordinates the hypersurfaces are:

$$p_1 = p_2^2, \qquad q_2 = 0$$

(compare with corollary 1 above). In the analytic or holomorphic case the normalising series (first found by Sato [19]) diverge, in general (Oshima [20]). The reason for this divergence is the presence of a 'functional module' in the holomorphic classification: the set of holomorphic equivalence classes of pairs of C^∞-equivalent holomorphic hypersurfaces is parametrised by a space of holomorphic functions intrinsically related to the pairs.

We will return to the classification of these pairs in chapter 7.

Excercise 1. A complete *flag* in a vector space V is a sequence of vector subspaces $V_0 \subset V_1 \subset \cdots \subset V_N = V$, $\dim V_i = i$. Classify the flags in a symplectic vector space V_{2n} up to linear symplectomorphisms.

Hint. Project V_{2n-1} to its characteristic space B_{2n-2} and proceed by induction.

Answer. The number of inequivalent flags is equal to $(2n-1)!!$. Each flag is equivalent to a coordinate flag $\{V_i = \mathrm{span}\,(e_1, \ldots, e_i)\}$, where $(e_1, \ldots, e_{2n})$ is a permutation of the *Darboux basis* $(\partial/\partial p_1, \ldots, \partial/\partial q_n)$. To list the representatives, put p_1 at the leftmost position, q_1 at a free place, p_2 at the leftmost free place, q_2 at a free place, etc. The generic flags correspond to the order $(p_1, q_1, \ldots, p_n, q_n)$. The codimension of the set of flags equivalent to one representative is equal to the number of transpositions in the permutation.

Example. The 3 types of flags in $\mathbf{R}^4$ are represented by the sequences

$$p_1 q_1 p_2 q_2, \qquad p_1 p_2 q_1 q_2, \qquad p_1 p_2 q_2 q_1.$$

The codimensions are 0, 1, 2, respectively.

Remark. The sum of the kernels of the restrictions of the symplectic structure to the spaces of a complete flag is always a Lagrangian subspace (M. Vergne, [21]).

Excercise 2. A (complete) *flag of submanifolds* of a manifold M is a sequence of submanifolds $M_0 \subset M_1 \subset \cdots \subset M_N = M$, $\dim M_i = i$. A flag of submanifolds of a symplectic space is called a *constant rank flag* if the rank of the restriction of the symplectic structure to any M_i is constant along M_i.

Classify the germs of constant rank flags of submanifolds of a symplectic manifold, up to symplectomorphisms.

Hint. Project M_{2n-1} to its characteristic space B_{2n-2} and use the known (by induction) structure of the latter to reduce to normal form the situation inside M_{2n-1} (inclusive the restriction of the symplectic structure). Then use Givental's theorem to rectify M_{2n-1} in M.

Answer. A germ of a constant rank flag is equivalent to a flag of vector spaces in a symplectic vector space. Hence the number of inequivalent germs in a symplectic $2n$-manifold is $(2n-1)!!$.

1.3 Lagrangian manifolds, fibrations, mappings, and singularities

Definition 1. A *Lagrangian submanifold* of a symplectic manifold is a submanifold of maximal dimension on which the symplectic structure vanishes.

(Its dimension is equal to one-half of the dimension of the symplectic manifold.)

Example 1. The plane $p = 0$ in $\mathbf{R}^{2n}$, equipped with Darboux coordinates, is a Lagrangian submanifold. All Lagrangian submanifolds of given dimension are locally symplectomorphic (by Givental's theorem), and hence each of them is locally defined by the equation $p = 0$ in certain Darboux coordinates.

Example 2. The formula $p = ds$, where s is a smooth function on V, defines a Lagrangian section of T^*V. For instance, the zero section of the cotangent bundle is a Lagrangian submanifold. A neighborhood of a Lagrangian submanifold is symplectomorphic to a neighborhood of the zero section of its cotangent bundle (this follows from Weinstein's theorem, see § 1.2).

Example 3. The fibers of the cotangent bundle are Lagrangian subspaces.

Example 4. Let Γ be a submanifold of a euclidean space $\mathbf{R}^n$ (Γ may be a hypersurface, a point, a curve,...). The set of oriented normals to Γ is a Lagrangian submanifold of the symplectic space of oriented lines in $\mathbf{R}^n$.

Example 5. The set of monic polynomials of degree $2m$ that are divisible by x^m is a Lagrangian subspace.

Remark. A. Weinstein's empirical law holds: in symplectic geometry *everything* is a Lagrangian submanifold (for instance, Hamilton equations and symplectomorphisms may be viewed as Lagrangian manifolds).

Definition 2. A *Lagrangian fibration* of a symplectic manifold is a fibration with Lagrangian fibers.

Example 1. $\mathbf{R}^{2n} \twoheadrightarrow \mathbf{R}^n$, $(q, p) \mapsto q$.

Example 2. The cotangent fibration over a smooth manifold is a Lagrangian fibration (the Lagrangian fibration of the phase space over the configuration space).

Example 3. Associate with any oriented line of a euclidean space its unit vector. We thus obtain a map from the symplectic manifold of oriented lines of the euclidean space to the unit sphere.

This map is a Lagrangian fibration (in essence it coincides with the cotangent bundle of the sphere, the only difference being the sign of the symplectic structure).

Theorem. *All Lagrangian fibrations (of fixed dimension) are locally equivalent (i.e. in suitable Darboux coordinates, any of them is defined by the formula in example 1 in a neighborhood of any point of the total space of the fibration).*

Proof. Let $\pi : E^{2n} \twoheadrightarrow B^n$ be a Lagrangian fibration and $f : B^n \to \mathbf{R}$ a function. Lift f to a function $F = f\pi : E^{2n} \to \mathbf{R}$ and consider the Hamiltonian flow g^t defined by the Hamiltonian vector field V_F (namely, $\omega(V_F, \xi) = dF(\xi)$ for any vector $\xi \in TE$, where ω is the symplectic structure of E).

Lemma. *The vector field V_F is vertical (tangent to the fibers), and the flows associated with different functions f, g on the base space commute.*

Proof. $dF(\xi) = 0$ for all vertical ξ. Hence V_F is skeworthogonal to the tangent space of the fiber. But this linear space is Lagrangian, hence its own skeworthogonal complement. Thus V_F belongs to this vertical Lagrangian space.

Commutativity follows, since the Poisson bracket $(F, G) = \omega(V_F, V_G) = 0$ for $F = f\pi$, $G = g\pi$.

In order to deduce the theorem from the lemma, we choose local coordinates $f_1, \ldots, f_n$ in the base and a local Lagrangian section $s : B \to E$ through the given point of E. Let $F_i = f_i\pi : E \to \mathbf{R}$ and let g_i^t be the corresponding flow. The point with Darboux coordinates (p, q) is $g_1^{-p_1} \cdots g_n^{-p_n} s(b)$, where b is the base point with $f_i(b) = q_i$.

Remark. A Lagrangian fibration intrinsically defines an affine structure on its fibers:

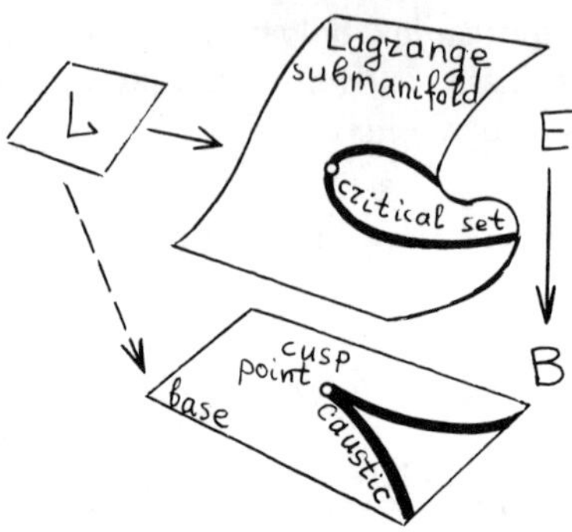

Figure 13: A Lagrangian map and its caustic

translations are defined by the flows generated by the Hamilton functions, lifted from the base manifold.

Example 4. Let an integrable Liouville system with integrals $I_1, \ldots, I_n$ have a compact regular integral n-manifold $I_1 = C_1, \ldots, I_n = C_n$, $(I_i, I_j) = 0$. In a neighborhood of this manifold the map I is a Lagrangian fibration. Hence the invariant tori of integrable systems form Lagrangian fibrations.

An affine structure in the fibers is the main ingredient in the construction of action–angle coordinates in integrable systems.

Definition 3 (*Lagrangian maps*). Consider an immersed Lagrangian submanifold L in the space of a Lagrangian fibration $E \twoheadrightarrow B$. The projection of L to B is called a *Lagrangian map*. Thus, a Lagrangian map is a triple $L \hookrightarrow E \twoheadrightarrow B$, where the left arrow is a Lagrangian immersion and the right one a Lagrangian fibration (Fig. 13).

A *Lagrangian equivalence* of two Lagrangian maps is a symplectomorphism of the total space transforming the first Lagrangian fibration to the second, and the first Lagrangian immersion to the second. Thus, a Lagrangian equivalence is a commutative (3×2) diagram; the two long horizontal lines are the given Lagrangian maps, the vertical lines are diffeomorphisms, and the middle line respects the symplectic structures.

The set of critical values of a Lagrangian map is called its *caustic*. Caustics of equivalent maps are diffeomorphic.

Example 1 (*gradient map*). $q \mapsto p = \partial S / \partial q$. The Lagrangian submanifold L is the graph of this map; the fibration is the cotangent bundle, $(q, p) \mapsto q$.

Example 2 (*normal map*). Associate with every vector normal to a submanifold its endpoint. This is a Lagrangian map (the Lagrangian submanifold L of $T^* \mathbf{R}^n$ is formed by the 1-forms $(n, \cdot)$ at the endpoints of the normal vectors n.

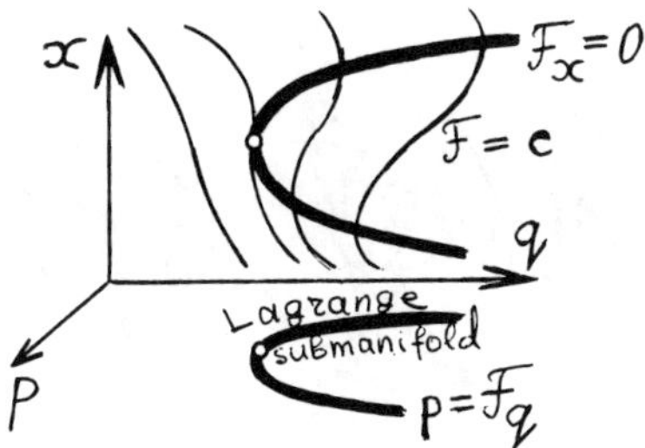

Figure 14: The Lagrangian curve generated by the family $x^3 - qx$

Its caustic is the envelope of the family of normal vectors to the initial submanifold. (For a hypersurface the caustic is also called the focal set of the hypersurface.).

Example 3 (*Gauss map*). This is a map from a transversally oriented hypersurface in a euclidean space to the unit sphere associating with a point on the hypersurface the unit normal vector at this point.

The Gauss map is a Lagrangian map. The Lagrangian submanifold of the symplectic manifold of oriented lines in a euclidean space is formed by the normals to the hypersurface.

These three examples have motivated (in 1966) the foundation of the theory of Lagrangian singularities, by analogy with the general singularity theory of Whitney—Thom—Mather.

Definition 4 (*Lagrangian singularities*). A *Lagrangian singularity* is a germ of a Lagrangian map, considered up to equivalence.

One important physical situation in which one encounters Lagrangian singularities is the problem of light caustics.

Let $F(x, q)$ be the optical distance from a point x of a (coherent) *source manifold* to the point q of the *observation manifold*. The phases of the waves on the observation manifold are defined (by Fresnel's principle of the stationary phase) by the Lagrangian manifold (Fig. 14)

$$L = \left\{ (p, q) : \exists x : \frac{\partial F}{\partial x} = 0, \quad p = \frac{\partial F}{\partial q} \right\}. \tag{$*$}$$

The family F of functions of x and depending on the parameter q is called the *generating family* of this Lagrangian submanifold (and of its Lagrangian map $(q, p) \mapsto q$ onto the observation manifold).

The caustics of the Lagrangian map are formed by the spots of unusual brightness.

Example. (See [2].) The Lagrangian singularities of generic Lagrangian maps of

Figure 15: The typical singularities of caustics in 3-space

Lagrangian manifolds of dimension $n \leq 5$ are, up to Lagrangian equivalence, contained in the following *list of Lagrangian singularities*, defined by the generating families ($\mu \leq n+1$):

A_μ: $F = \pm x^{\mu+1} + q_1 x^{\mu-1} + \cdots + q_{\mu-1} x$;

D_μ: $F = x_1^2 x_2 \pm x_2^{\mu-1} + q_1 x_2^{\mu-2} + \cdots + q_{\mu-2} x_2 + q_{\mu-1} x_1$;

E_6: $F = x_1^3 \pm x_2^4 + q_1 x_1 x_2^2 + q_2 x_1 x_2 + q_3 x_2^2 + q_4 x_1 + q_5 x_2^2$;

E_7: $F = x_1^3 + x_1 x_2^3 + q_1 x_1^2 x_2 + q_2 x_1^2 + q_3 x_1 x_2 + q_4 x_2^2 + q_5 x_1 + q_6 x_2$;

E_8: $F = x_1^3 + x_2^5 + q_1 x_1 x_2^3 + q_2 x_1 x_2^2 + q_3 x_2^3 + q_4 x_1 x_2 + q_5 x_2^2 + q_6 x_1 + q_7 x_2$.

All singularities A, D, E defined by these generating families (for arbitrary n) are stable and simple (have no moduli). The simplest singularities A_2 (fold) and A_3 (cusp) are explicitly given by the projections $(q, p) \mapsto q$ of the Lagrangian manifolds

A_2: $q_1 = \mp 3x^2$, $p_1 = x$;

A_3: $q_2 = \mp 4x^3 - 2xq_1$, $p_1 = x^2$, $p_2 = x$.

Both folds ($\mp$ in A_2) are Lagrangian equivalent, but the two types of Lagrangian cusps ($\mp$ in A_3) are not.

Thus, generic Lagrangian 2-surfaces in a phase 4-space define, under projection to the configuration 2-plane, the same Whitney singularities as generic (non-Lagrangian) 2-surfaces (up to diffeomorphism). This is not *a priori* evident, since Lagrangian maps are rather special. Indeed, in higher dimensions there are differences between generic general maps and generic Lagrangian maps: some generic general singularities are not encountered in the list of generic Lagrangian singularities, while some generic Lagrangian singularities are not generic (non-Lagrangian) general singularities.

Example 1. A generic 1-dimensional caustic has (besides selfintersections) only ordinary cusp singularities (A_3). A generic 2-dimensional caustic has (besides selfintersections) only swallowtails (A_4), pyramids (D_4^-), and purses (D_4^+), Fig. 15. (These D singularities were called 'umbilical singularities' by R. Thom [22] because of their relation with umbilical points of 2-surfaces in euclidean 3-space; they are the singularities of focal sets.)

Example 2. A Gaussian map of a generic surface in euclidean 3-space has only *folds* (A_2) (at generic points of a parabolic line) and *cusps* (A_3) (at the special points of the parabolic line at which the asymptotic direction is tangent to this line). See [23]–[26] for more details.

The generic Lagrangian singularities of maps of spaces of dimension $n > 5$ have moduli (continuous invariants for Lagrangian equivalence), while for higher dimensions also functional moduli appear. Despite this difficulty, there exists a classification, up to Lagrangian equivalence, of the generic Lagrangian singularities for $n \leq 10$ (see [27]–[29]). The corresponding normal forms contain arbitrary parameters (moduli) and functions (functional moduli).

This classification is finer than the topological classification (up to homeomorphism). It is known that the number of topologically distinct types of Lagrangian singularities of generic Lagrangian maps is finite for any dimension n ([30], [31]). However, for $n \leq 10$ it is a difficult problem to explicitly find the topological classification, although the smooth classification is known explicitly. The difficult point is to find the values of the moduli for which the corresponding singularity is topologically inequivalent to the singularity corresponding to generic nearby values of the moduli. Such exceptional moduli values exist, but their computation (even in the simplest cases) seems to be beyond the possibilities of modern computers.

Let us return to the general theory of Lagrangian singularities. Every Lagrangian singularity can be defined by a generating family F of functions of variables x and parameters q. (Such a family defines a smooth Lagrangian manifold if the equation $\partial F / \partial x = 0$ satisfies the conditions of the implicit function theorem; this transversality condition is included in the definition of generating family.)

Two families of functions are called R^+-*equivalent* if each of them can be transferred into the other by a smooth change of the independent variables (smoothly depending on the parameters), a smooth change of the parameters, and by adding a smooth function of the parameters; in outdated notation:

$$F(x, q) \sim F(X(x, q), Q(q)) + \Phi(q).$$

Two families of functions (with argument spaces of possibly different dimensions) are called R^+-*stably equivalent* if they become R^+-equivalent when nondegenerate quadratic forms in new variables are added to them.

Example. The function x^3 of one variable x is stably equivalent to the function $x^3 - y^2$ of two arguments (x, y), but it is not stably equivalent to the function x^3 regarded as a function of these same two arguments.

Theorem. *All generating families of Lagrangian equivalent singularities are locally stably R^+-equivalent. Stably R^+-equivalent generating families define Lagrangian equivalent maps.*

For the proof of this theorem see, for instance, [2], [28].

Thus, the classification of Lagrangian singularities (up to Lagrangian equivalence) is reduced to the classification of the germs of families of functions (up to R^+-stable equivalence).

For instance, Lagrangian stability means *versality* of the corresponding family [28].

The well-known *hierarchy of critical points* of smooth functions (see [2])

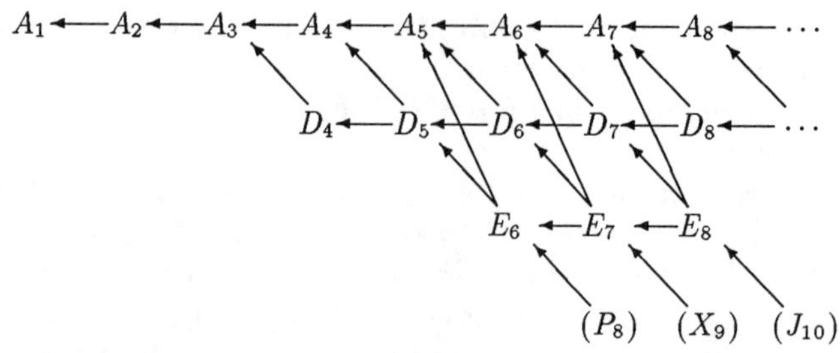

implies a hierarchy of Lagrangian singularities, etc.

Chapter 2

Applications of the theory of Lagrangian singularities

We will give five applications of the classification of Lagrangian singularities: to the method of the stationary phase (that is, to the theory of integrals of rapidly oscillating functions), to the asymptotics of the number of (integer) lattice points in generic large smooth domains, to the cosmological theory of the large scale structure of the Universe, and to the study of perestroikas of optical caustics and shock waves.

2.1 Oscillatory integrals

The caustic of a Lagrangian map can be described as the set of points q at which the following integral (of a rapidly oscillating function) decreases slower than at generic points q (as the wavelenght h decreases):

$$I(h,q) = \int e^{iF(x,q)/h} a(x,q)\, dx.$$

The real valued function F of the variables x and parameters q is called the *phase*, and a is called the *amplitude* of the integral. The dependence of the integral on the amplitude is linear. In most cases the amplitude is supposed to be infinitely smooth and to vanish outside a compact set. When discussing asymptotics of the integral, we mean the comportment of the corresponding linear functional of the amplitude function. Thus the asymptotics of the integral is understood to be its behavior for the 'most worse' amplitude (that is, for a smooth amplitude that gives slowest decrease of the integral as $h \to \infty$), and hence for any generic amplitude.

The simplest example of an oscillatory integral is a *Fresnel integral* $\int \cos(x^2/h)\, dx$. The 'main' part of this integral comes from a small neighborhood of the critical point of the phase $F = x^2$ (Fig. 16). The diameter of the central bump of the function

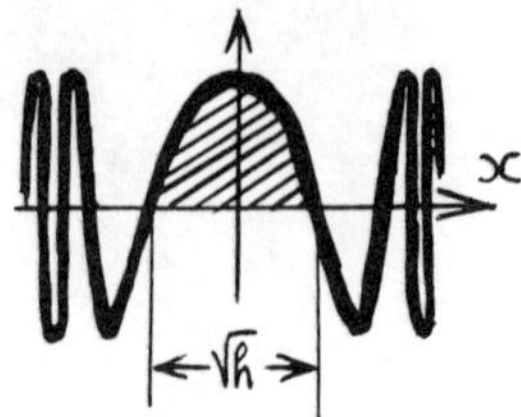

Figure 16: An oscillatory Fresnel integral with Morse phase function

$\cos(x^2/h)$ is of order $\sqrt{h}$. Hence the whole integral decreases like $h^{1/2}$ (the following positive and negative bumps compensate each other).

For n independent variables x_i and for nondegenerate *Morse critical points* of the phase function (that is, when $\det \partial^2 F / dx_i\, dx_j \neq 0$ at the critical points) the integral decreases like $h^{n/2}$. (For the proof one locally reduces F to a sum of squares, and then uses the Fresnel integral of Fig. 16.)

The coefficient of $h^{n/2}$ in the asymptotics is proportional to $|\det(\partial^2 F / dx_i\, dx_j)|^{-1/2}$ (this factor comes from the Jacobian of the change of variables reducing the phase to a sum of squares). Colin de Verdière [32] has proved the following upper bound for the integral in terms of the sum of the asymptotics at the complex Morse critical points of a holomorphic phase function:

$$|I| \leq C h^{n/2} \sum \left| \det \left(\frac{\partial^2 F}{\partial x_i \partial x_j} \right) \right|^{-1/2} ;$$

this bound holds at the noncaustic points q, uniformly in a neighborhood of the caustic. It has been proved for phase functions depending generically on 6 or fewer parameters (or for a phase function with only simple and parabolic Lagrangian singularities). However, I do not know a counterexample for the general case (of generic families with a larger number of parameters).

At a point q belonging to the caustic, the corresponding phase function $F(\cdot, q)$ has, as a function of x, a degenerate (nonMorse) critical point. For such a q the integral decreases more slowly, and the index $\beta(q)$ of the main term of the asymptotic expansion

$$I(h, q) \sim C h^{\beta(q)} \ln^{\gamma(q)} h + \cdots, \qquad h \to 0,$$

is smaller than the usual value $n/2$.

The indices $\beta(q)$ and $\gamma(q)$ can, in many cases, be found from the Newton boundary of the phase function (Fig. 17). Consider the Taylor series of the phase function at a critical point. Bringing the critical point to the origin, we obtain the Taylor series $F \sim \sum a_m x^m$ (where the value of the parameter q is fixed), The index m belongs to the positive octant of the integer lattice $\mathbf{Z}^n$. The *support* of (the series of) F is the set of those m for which the corresponding coefficient a_m is nonzero. Adjoin to every point of the support the complete nonnegative octant $\mathbf{R}^n_+$. The boundary of the convex hull of the 'staircase' thus obtained (i.e. of the union of the octants $m + \mathbf{R}^n_+$) is called the *Newton boundary* of (the series of) F.

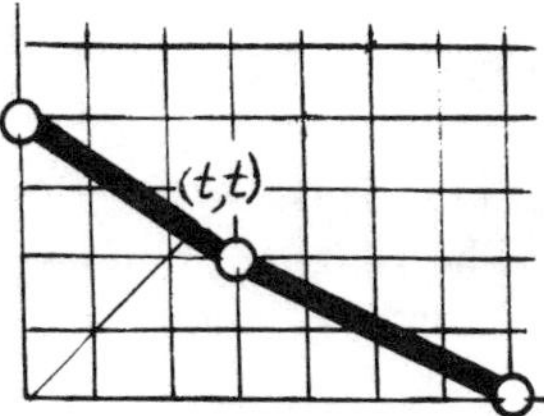

Figure 17: The Newton boundary and its index $\beta = 1/t$

A remarkable fact, discovered by Newton, is that, at least for generic values of the coefficients, most topological or discrete invariants of a critical point do not depend on the values of the coefficients of the series, and do not depend on the terms outside the Newton boundary (which is a multi-dimensional generalisation of the order of a zero of a function of one variable).

This general principle leads to a lot of interesting formulas, giving the values of topological and other discrete invariants of singularities (Hodge numbers, spectra,...) in terms of the geometry of integer convex polyhedra (Newton boundaries). Many interesting formulas of this kind can be found in [33]–[35]. It is interesting to note that this relation between the topology of singularities and the geometry of convex bodies is useful in both directions, since one can use relations between invariants of singularities (known from topology, algebraic geometry, etc.) to obtain highly nontrivial theorems on the combinatorics of convex polyhedra. (The proof by Hovanski and Prohorov on the absence in Lobachevski spaces of dimension exceeding 955 of reflection groups with fundamental domains of finite volume is just one example.)

Returning to the asymptotics of oscillatory integrals, we find the following formula for the index β of the main term (conjectured by the author and proved by Varchenko).

Consider the point $(t, t, \ldots, t)$ of intersection of the Newton boundary with the diagonal of the positive octant. The *index β of the Newton boundary* is, by definition, $1/t$.

Example 1. The index β of the Newton boundary of the Morse function $x_1^2 + \cdots + x_n^2$ is equal to $n/2$.

The *index γ of a Newton boundary* is, by definition, the dimension of the set of supporting hyperplanes of this boundary at the point $(t, t, \ldots, t)$.

The boundary is *far* (from the origin) if $t > 1$.

Theorem (A.N. Varchenko [33]). *1) For a generic phase function with given Newton boundary the indices β and γ of the main term of the asymptotics are equal to the indices β and γ of the Newton boundary, provided that the boundary is far (from the origin).*

2) When the boundary is not far, β(generic function)$\geq$ β(boundary).

3) For analytic phase functions depending on two variables,

$$\beta(generic\ function) = \beta(boundary),$$

in a suitable analytic coordinate system.

In this theorem, 'nongeneric' functions are functions for which the Taylor coefficients of the terms belonging to the Newton boundary satisfy a nontrivial algebraic equation. If the boundary if far (from the origin), then $\beta(F) \leq \beta(\text{boundary})$ for *any* function with that boundary.

In other words, *if the boundary is far (from the origin), the oscillatory integral never decreases faster than the boundary prescribes, and its main part is almost always exactly as the boundary prescribes.*

In case the boundary is not far, a generic integral decreases at least as rapidly as the boundary prescribes. In this case the theorem leaves open the possibility of stably faster decrease. This means that in this case the integral behaves like in the situation of a singularity that is weaker than the boundary prescribes. The explanation is that the strong singularity prescribed by the boundary may be a purely complex phenomenon, invisible in the real domain. Varchenko's theorem implies that such 'purely complex phenomena' are impossible if the boundary is far.

Varchenko has found an example of a phase function for which the 'purely complex phenomenon' does occur. The corresponding Newton boundary in the 5-space is not far from the origin: $\beta(\text{boundary})= 29/18$, while $\beta(F) = 7/4 > 29/18$ for an open set of phase functions having this Newton boundary. Namely, the phenomnon occurs for the Newton boundary of the function

$$u^2 + x^9 + y^9 + z^9 + (u - (x^2 + x^4 + y^2 + z^2))v.$$

For functions of $n \leq 3$ variables such an example is impossible, and the condition on the Newton boundary in the first assertion in the theorem concerning β can be discarded (the index $\gamma(F)$ may be equal to $\gamma(\text{boundary})-1$ for certain phase functions of 3 variables; these numbers are equal also for the function $F = xy$ of 2 variables).

We return to an oscillatory integral with phase $F(x, q)$ depending on the parameter q. For noncaustic values of q the integral decreases like $h^{n/2}$ as $h \to 0$, where n is the dimension of $\{x\}$. At generic nonsingular points of a caustic the integral decreases more slowly, like $h^{n/2-1/6}$.

For values of q belonging to the singular part of the caustic the integral generically decreases even slower as $h \to 0$. The worse the singularity, the slower the decrease. In the simplest cases the integral can be majorised (uniformly with respect to q in a neighborhood of a given q_0) by the main term of its asymptotics at q_0, or at least by $Ch^{\beta(q)-\epsilon}$, for any $\epsilon > 0$.

For $n = \dim\{x\} = 1$ (and smooth amplitudes a vanishing outside a compact set) such a uniform bound (with $\epsilon = 0$) was proved by I.M. Vinogradov [37]. For $n = 2$ it was proved by V.N. Karpouchkin [38], [39].

For $n = 3$ Varchenko has constructed a counterexample; in it β is not semicontinuous and hence the uniform bound does not hold. Namely, for

$$F = (qx_1^2 + x_1^4 + x_2^2 + x_3^2)^2 + x_1^{12} + x_2^{12} + x_3^{12}$$

the index jumps at $q = 0$:

$$\beta(q = 0) = \frac{5}{8}, \qquad \beta(q < 0) = \frac{7}{12} < \frac{5}{8}.$$

The phenomenon of nonsemicontinuity is unavoidable in families depending on $l \geq 73$ parameters. If the number of parameters is small, the bound holds generically (for instance, V.N. Karpouchkin has proved it for $l \leq 7$). The exact value of l for which the uniform bound first fails to hold is unknown.

F. Pham has conjectured [40] that it holds generically for families with arbitrary many parameters, provided that the definition of β is changed so as to take into account complex chains (considering integrals for which the saddle point method applies, rather than oscillatory integrals). The hierarchy of exponential integrals with many confluent saddle points has been studied by V.A. Vassiliev [41]. Pham's conjecture follows from the more general theory developed by A.N. Varchenko.

Asymptotic series for integrals (with arbitrary amplitudes a and integration chains) have been used by Varchenko ([42]–[44]) to define the mixed Hodge structures of singularities. A distinct mixed Hodge structure was defined earlier by J.Steenbrink [45] (the discrete invariants—the Hodge numbers—of both structures coincide, but the filtrations that they define in the cohomology of the Milnor fibers are distinct). The theory of mixed Hodge structures of singularities implies many important results. Among them are the proof by Varchenko and Steenbrink of old conjectures concerning semicontinuity of the singularity spectrum [46], [47], concerning constancy of β along a stratum $\mu = \text{const}$ [48], and concerning the coincidence of the modality and the so-called inner modality of quasihomogeneous functions [49].

The mixed Hodge structure of a singularity implicitly contains a lot of geometric information; however, the information is well hidden.

Consider the Poincaré index of the gradient vector field of a real valued smooth function of $2m$ variables at a critical point. Ths index is majorised by the middle Hodge number of the mixed Hodge structure associated with the eigenvalue 1 of the monodromy operator [34],

$$|\text{ind}| \leq h_1^{m,m}.$$

The mixed Hodge structure of a singularity provides a natural way of defining the arguments of the eigenvalues of the monodromy operator, $e^{2\pi i p/q}$. For instance, for Pham— Brieskorn singularities,

$$f(x) = x_1^{a_1} + \cdots + x_n^{a_n},$$

the spectrum is the set of rational numbers ($\mu = (a_1 - 1) \cdots (a_n - 1)$ in total)

$$\frac{p}{q} = \frac{k_1}{a_1} + \cdots + \frac{k_n}{a_n}, \qquad 0 < k_j < a_j, \quad 1 \leq j \leq n.$$

The condition of semicontinuity of the spectrum is similar to the Fisher—Rayleigh—Courant principle, according to which the lengths of the axes of an ellipsoid are separated by the length of the axis of any hyperplane section of it. Consider a singularity S' adjacent to a given (more singular) singularity S. Let the spectra be $\alpha_1 \leq \alpha_2 \leq \ldots \leq \alpha_\mu$ for S and $\alpha'_1 \leq \alpha'_2 \leq \ldots \leq \alpha'_{\mu'}$ for S' ($\mu' < \mu$) (both spectra are symmetric with respect to $n/2$). The semicontinuity property is the set of inequalities

$$\alpha_1 \leq \alpha'_1,\ \alpha_2 \leq \alpha'_2, \ldots,\ \alpha_{\mu'} \leq \alpha'_{\mu'}.$$

If $\mu' = \mu - 1$, this is equivalent to the separation property

$$\alpha_1 \leq \alpha'_1 \leq \alpha_2 \leq \alpha'_2 \leq \ldots \leq \alpha'_{\mu'} \leq \alpha_\mu.$$

Remark 1. One may think of the spectrum (possibly differently scaled) as being the set of eigenfrequencies of an oscillator associated with the singularity (possibly the quasiclassical asymptotics of part of the spectrum of a quantum system, this part being separated from the infinite part by a spectral gap, like for the Fokker—Planck equation on a closed manifold).

Thus we may consider semicontinuity as a manifestation of the existence of a hidden oscillatory system, formed by the vanishing cycles of the singularity.

Remark 2. Experimental material shows that adjacent singularities of critical points of functions are generated by codimension-1 adjacencies: for any pair of adjacent singularities $S' \leftarrow S$ there exists a chain

$$S' \leftarrow S_0 \leftarrow S_1 \leftarrow \cdots \leftarrow S_r = S,$$

where $\mu(S_i) - \mu(S_{i-1}) = 1$. It is unknown whether this experimental rule holds in general.

Remark 3. Many integer numbers associated to a singularity should be majorised by invariants of the mixed Hodge structures and/or by numerical characteristics of the Newton boundary. For instance, the Morsifications of a given singularity have Morse numbers M_i (of nondegenerate critical points of arbitrary fixed negative index of inertia i of the second differential) and Betti numbers b_i (of real level hypersurfaces). The maxima of such quantities as M_i, $M_0 - M_1 + \cdots \pm M_i$; b_i, $b_0 - b_1 + \cdots \pm b_i$ over all Morsifications of a given singularity are important topological characteristics of the singularity; finding exact bounds for them is an old problem in singularity theory [50].

However, little is known besides the direct generalisations of well-known results in real algebraic geometry [51]–[53], even in the simplest particular cases. For instance, consider a real polynomial of degree d in two variables. It can be easily seen that the sum of the numbers of minima and maxima points, $M_0 + M_2$, is majorised by $d^2/2 + O(d)$, but it is not known how large the difference $M_0 - M_2$ can be. For the product of d linear functions the asymptotic answer has been found by Ju. Chekanov, using mod p arithmetic of elliptic curves [54]: $M_0/M_2 \leq 2$ and for certain configurations of lines, $M_1/M_2 \geq 2 - O(1/d)$ (see also [55]–[57]).

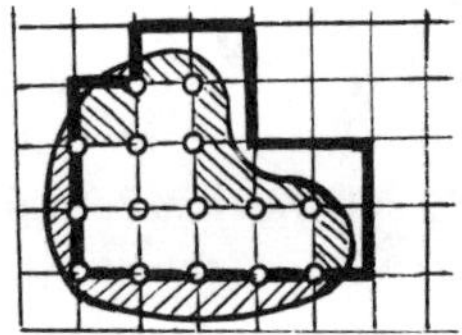

Figure 18: Lattice points in a dilated domain

One classical problem of this kind (related to complex singularities) is the determination of the maximal number ν of Morse critical points of a hypersurface of degree d in n variables. For $n = 2$, $\nu = d(d - 1)/2$ (it is attained for d lines). For $n = 3$ the maximal number of conical Morse singular points is known only for $d < 7$:

d	1	2	3	4	5	6	7
D	0	1	4	16	31	64?	$90 \leq \nu \leq 108$

Chmoutov's examples [44], based on the Tchebyshev polynomials $f = \sum T_d(x_i)$, provide for $n = 3$ the bound $\nu \geq \frac{3}{8}d^3 + O(d^2)$ for the number of Morse points. Miaoka's inequality [58] implies in this case $\nu \leq \frac{4}{9}d^3 + O(d^2)$. The exact coefficient in the upper bound seems to be unknown still (see [59] for more details on this problem and on related inequalities for the numbers of coexisting non-Morse singularities).

2.2 Lattice points

Let G be a compact domain in $\mathbf{R}^n$, bounded by a smooth hypersurface, and let V be its volume. By λG we denote a homothetical domain, and by $N(\lambda)$ the number of lattice points in λG. Consider the remainder in the asymptotic formula $N(\lambda) \sim \lambda^n V$, $\lambda \to \infty$:

$$R(\lambda) = \lambda^n V - N(\lambda).$$

The modern theory of Lagrangian singularities and oscillatory integrals leads to interesting relations between differential geometry and number theory.

1°. It is clear that $|R(\lambda)| \leq C\lambda^{n-1}$, since the remainder is majorised by the number of lattice points in a d-neighborhood of the boundary of the domain λG, where d is the diameter of the unit cube, and the area of the boundary scales like λ^{n-1} (Fig. 18).

2°. The remainder cannot be smaller than the number of lattice points on the boundary. Indeed, an arbitrary small variation of λ leads to variation in $N(\lambda)$ equal to the number of lattice points on the boundary. Hence, for a ball centered at the origin, $|R(\lambda)| \geq C\lambda^{n-2}$ for certain (arbitrary large) values of λ.

To see this, consider the layer between the spheres of radii r and $r+a$. The number of lattice points in this layer scales like r^{n-1}. The number of spheres containing these points (that is, the number of integers between r^2 and $(r+a)^2$) scales like r. Hence, by Dirichlet's box principle, there is a sphere that contains at least cr^{n-2} lattice points. It follows that $|R(\lambda)| \geq C\lambda^{n-2}$ for certain (arbitrary large) values of λ.

$3°$. Colin de Verdière [32] has deduced from the theory of Lagrangian singularities the upper bound

$$|R(\lambda)| \leq C\lambda^{n-2+\frac{2}{n+1}}$$

for domains G bounded by generic hypersurfaces in $\mathbf{R}^n$, where $n \leq 7$.

It is not known whether this bound holds in higher dimensions. For $n = 2$ the index in this bound is best possible; for n larger the index might be lower, but this has not been proved, even not in the case of positive curvature.

$4°$. The remainder R can be regarded as being the sum of many ($\sim \lambda^{n-1}$) summands centered at 0 (each summand corresponds to a cubic cell cut by the boundary hypersurface). If these summands were independent random variables, then (by the laws of probability theory) the sum should scale like $\lambda^{(n-1)/2} \leq \lambda^{n-2}$.

The theory of Lagrangian singularities provides some substantiation for this heuristic reasoning.

Theorem (A.N. Varchenko [60]). *The quadratic mean value of the remainder $|R(\lambda)|$, averaged over the set of of lattices obtained from the integer point lattice by translations and rotations, is majorised by $C\lambda^{(n-1)/2}$ for any domain in $\mathbf{R}^n$ with smooth boundary.*

The convex analytic case was studied by Randol [61].

The idea of the proof is similar as that in the proof of the unitarity of Maslov's 'canonical operator': the phase function $f(x) - (q, x)$, considered as a function of both arguments x and q, has only Morse critical points.

$5°$. The relation between the problem of counting lattice points and the theory of oscillatory integrals depends on the theory of mechanical quadrature formulas.

In order to estimate the error of a quadrature formula for a function on the torus it is natural to first apply the formula to harmonics (and subsequently representing a function as a linear combination of harmonics). For harmonics the error can be explicitly computed: it equals 0, except for certain special harmonics with large wave vectors (proportional to λ). Hence the error depends mainly on the Fourier coefficients with such large wave vectors of the characteristic function of the set G (that is, of the function equal to 1 inside G and vanishing elsewhere in the lattice cube).

These coefficients are also the values of oscillatory integrals along the boundary of the domain (the phases of these integrals are the restrictions to the boundary of the linear functions which are the phases of the harmonics).

Thus, the error can be measured in terms of oscillatory integrals whose critical points of the phase reflect the inflections and flattening of the boundary. Hence the singularities of the system of rays orthogonal to the boundary govern the asymptotics of the number of lattice points in dilated domains.

The remainder $R(\lambda)$ can be written as

$$\frac{R(\lambda)}{\lambda^n} = \int f\,dx - \sum \frac{f(z/\lambda)}{\lambda^n}, \qquad z \in \mathbf{Z}^n, \tag{$*$}$$

representing the error of a quadrature formula with step $1/\lambda$ applied to the characteristic function f of G. Assume, for the sake of simplicity, that G lies in the unit cube, and periodically extend the characteristic function to the whole of $\mathbf{R}^n$. Let us also assume that λ is an integer (these restrictions can be easily eliminated).

For a periodic function, in formula (*) the integral is understood to be along one period-cube while the sum extends over the λ^n summands corresponding to the points of this cube (in other words, we integrate over the torus $\mathbf{R}^n/\mathbf{Z}^n$ and sum over the subgroup of roots of the identity of order λ on this torus).

Denote the harmonics (the characters of the torus) by $e_k = e^{2\pi i(k,x)}$, where k is a point of the dual lattice $\mathbf{Z}^{n*}$.

For $f = e_k$ the error of the quadrature formula is:

$$\left(\frac{\sum}{\lambda^n} - \int\right) e_k = \begin{cases} 1 & \text{if } k \in \mathbf{Z}^{n*}\backslash 0, \\ 0 & \text{otherwise.} \end{cases}$$

For an f with Fourier series $\sum f_k e_k$ we find

$$\left(\frac{\sum}{\lambda^n} - \int\right) f = \sum f_k, \qquad k \in \mathbf{Z}^{n*}\backslash 0.$$

The wave vectors of the harmonics in the sum are multiples of λ, and hence are large for λ large ($|k| \geq \lambda$). Also, the Fourier coefficient f_k equals the integral over a period:

$$f_k = \int f(x)e^{-2\pi i(x,k)}\,dx.$$

The exponential in this integral is rapidly oscillating for k large. If f is the characteristic function of G, the integral can be transformed to an integral over the boundary, with phase $-2\pi i(x,k)$. Degeneration of critical points of the restriction to the boundary of this linear function is responsible for the order of magnitude of the coefficients f_k, and hence for the order of magnitude of the remainder $R(\lambda)$.

Of course, the Fourier series of a characteristic function converges badly. Hence the derivation of a bound for R from bounds on the Fourier coefficients requires some smoothing of the characteristic function (the so-called van der Corput method). For the details of this technique see [32], [60], [61].

$6°$. We will mention one more estimate related to lattice points and discovered as a byproduct of the classification of Lagrangian singularities by their Newton polyhedra.

Theorem (G.E. Andrews [62], C. Sevastjanov, S. Konjagin [63]). *The number of vertices of a convex polyhedron in* $\mathbf{R}^n$ *of volume* V *and with integer vertices does not exceed* $CV^{(n-1)/(n+1)}$.

This same bound applies to the number of faces of any, fixed, dimension of such a polyhedron.

Examples showing that the order of this bound cannot be lowered are described in [64], in which the bound was proved for $n = 2$. The number of vertices of the planar Newton diagram of a singularity bounding an area S cannot exceed $CS^{1/3}(\ln S)^{2/3}$. The logarithm of the number of Newton diagrams bounding an area S is bounded from above and below by numbers proportional to $S^{1/3}(\ln S)^{2/3}$ (for the number of classes of closed convex polygons bounding an area S these bounds are proportional to $S^{1/3}$). All these results are due to Konjagin.

Remark 1. The above bounds on the entropy of the set of convex hypersurfaces with integer vertices can be interpreted as bounds on the influence of convexity and integer quantisation: the entropy of the set of integer convex surfaces scales like the ϵ-entropy of the set of functions of smoothness $1 + 1/n$. Since convexity is, more or less, equivalent to 1-smoothness, the price of quantisation is $1/n$th of one derivative.

Remark 2. The interaction of smooth and integer structures, crucial in lattice point problems, is equally important in other problems of calculus.

Consider the configuration consisting of a smooth submanifold of a euclidean space with respect to the hyperplanes defined by equations with integer coefficients. The problem of giving bounds on the distance of a typical point of a generic submanifold to hyperplanes defined by equations with bounded integer coefficients is studied in the theory of diophantine approximations on submanifolds [65], [66]. This (unsolved) problem is important in many applications, for instance in the study of resonance in nonlinear oscillations (see [17], chapt. 4). The answer depends on bounds on the curvature. Flattening of the boundary enhances phase locking at resonances.

For instance, the long time mean velocity of the drift of the action variables in a Hamiltonian system close to integrable has been bounded by the Nehoroshev exponential bound; it involves Nehoroshev numbers and depends on the type of curvature of the characteristics of the level sets of the Hamilton function. For analytic functions the Nehoroshev curvature condition can be simply formulated as follows: the critical points of the restrictions to affine subspaces (of arbitrary dimension) of the action space of the unperturbed Hamilton function should be isolated in the complex plane (Ju.S. Iljashenko, [67]).

The exact values of the Nehoroshev numbers have presently been computed only for generic systems with 2 or 3 degrees of freedom [68].

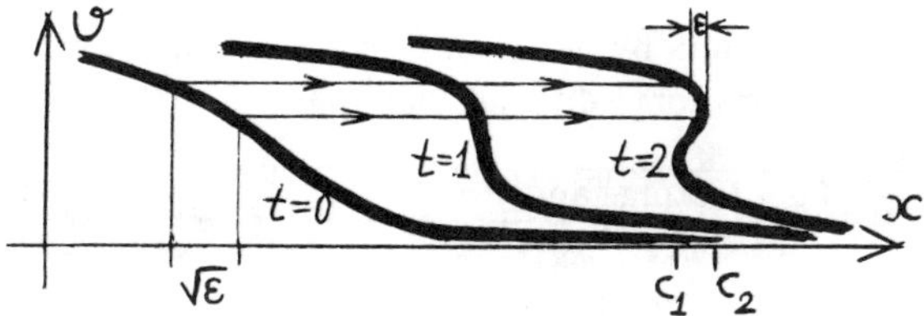

Figure 19: The velocity graph in a medium of noninteracting particles

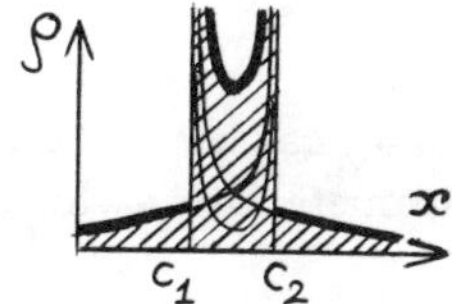

Figure 20: The density of the medium after the formation of a caustic

2.3 Perestroikas of caustics

In a generic 1-parameter family of Lagrangian maps there are nongeneric singularities (for certain parameter values). When the parameter passes through such a value, the caustic changes its shape. These perestroikas (metamorphoses, bifurcations) of caustics in generic families of low-dimensional spaces are described below.

Example. Inertial motion of a continuum of interacting particles may be described in terms of a 1-parameter family of maps $x \mapsto x + tv(x)$ (Fig. 19). In the case of a potential velocity field v these maps are Lagrangian maps. The caustics of such maps are the places of particle concentration (Fig. 20). According to Ja.B. Zeldovich [69], similar models (taking into account the gravitation and expansion of the Universe) describe the formation of inhomogeneities of the density (of the large scale structure of the Universe) at the early stage of expansion of a hot almost homogeneous Universe.

[The Lagrangian nature of the map is preserved for motion of particles in a potential force field (since the phase flow of any Hamiltonian system sends Lagrangian submanifolds to Lagrangian submanifolds). The field may be time-dependent or generated by the moving particles themselves (for instance, the Lagrangian nature is preserved under motion of particles in their own gravitational field).

However, after the first perestroika (i.e. after the formation of a caustic) a Lagrangian submanifold acquires certain singularities ('weak discontinuities'), since the density (and hence the Hamiltonian) becomes singular at the caustic.

It seems that these singularities of Lagrangian maps are too weak to destroy the topology of the caustics and of their perestroikas. However, this has not yet been proved rigorously, even not in the 1-dimensional case (see [70], [71]).]

The perestroikas of caustics can be described as the metamorphoses of sections of 'big caustics' in space–time by isochrones (see [1], [72]). The normal forms of big caustics and time functions for 3-dimensional space are represented in the following table:

type	generating family	time function
A_3	$x^4 + q_1 x^2 + q_2 x$	q_3 or $\pm q_1 \pm q_3^2 \pm q_4^2$
A_4	$x^5 + q_1 x^3 + q_2 x^2 + q_3 x$	q_4 or $\pm q_1 \pm q_4^2$
A_5	$x^6 + q_1 x^4 + q_2 x^3 + q_3 x^2 + q_4 x$	$\pm q_1$
$D_4^\pm$	$x_1^2 x_2 \pm x_2^3 + q_1 x_2^2 + q_2 x_2 + q_3 x_1$	q_4 or $\pm q_1 + a q_2 + q_3 \pm q_4$
D_5	$x_1^2 x_2 + x_2^4 + q_1 x_2^3 + q_2 x_2^2 + q_3 x_2 + q_4 x_1$	$\pm q_1 + q_4 + a q_2$

Here $(q_1, \ldots, q_4)$ are the coordinates in space–time and a is a real parameter.

We recall the definition of (big) caustic, in terms of the generating family $F(x, q)$: it is the set of values of the parameter q for which the corresponding function $F(\cdot, q)$ has a nonMorse critical point, i.e.

$$\{q : \exists x : F_x = 0, \det F_{xx} = 0\}.$$

This caustic is the set of critical values of the projection on the q-space along the p-space of the corresponding Lagrangian manifold

$$\{(p, q) : \exists x : F_x = 0, p = F_q\}.$$

The perestroikas corresponding to the families in the table are presented in Fig. 21. (They first appeared in [73]; for more details see [72], [74].) The parameter a is a modulus, hence there exists a continuous family of differentiably inequivalent perestroikas (while there are only 6 inequivalent germs of singularities of generic big caustics in 4-dimensional space–time).

The number of topologically inequivalent perestroikas is finite. In order to obtain the complete list of topological classes of generic perestroikas in physical 3-space it suffices to put $t = q_1 + q_2$ in the case of D_4^-, $t = q_1 \pm q_2$ or $t = q_1 + q_3$ in the case of D_4^+, and $t = \pm q_1$ in the case of D_5. Hence the number of distinct types is 11 (neglecting the orientation of the time axis).

Example. Consider the formation of the first caustic (of type A_3, $t = -q_1 + q_3^2 + q_4^2$). In order to understand what happens in the 3-dimensional case, it is convenient to begin with the 2-dimensional case (to put $q_4 = 0$ in the formulas).

The big caustic in 3-dimensional space–time (Fig. 22) is a surface with a cuspidal edge (in the space with coordinates (q_1, q_2, q_3) the equation of this surface

$$\{q : \exists x : F_x = F_{xx} = 0, F = x^4 + q_1 x^2 + q_2 x\}$$

is $8q_1^3 + 27q_2^2 = 0$).

The isochrones $t = $ const do not intersect the big caustic if $t < 0$. For $t = 0$ a point caustic appears. It starts growing immediately (infinitely rapidly, from the very first

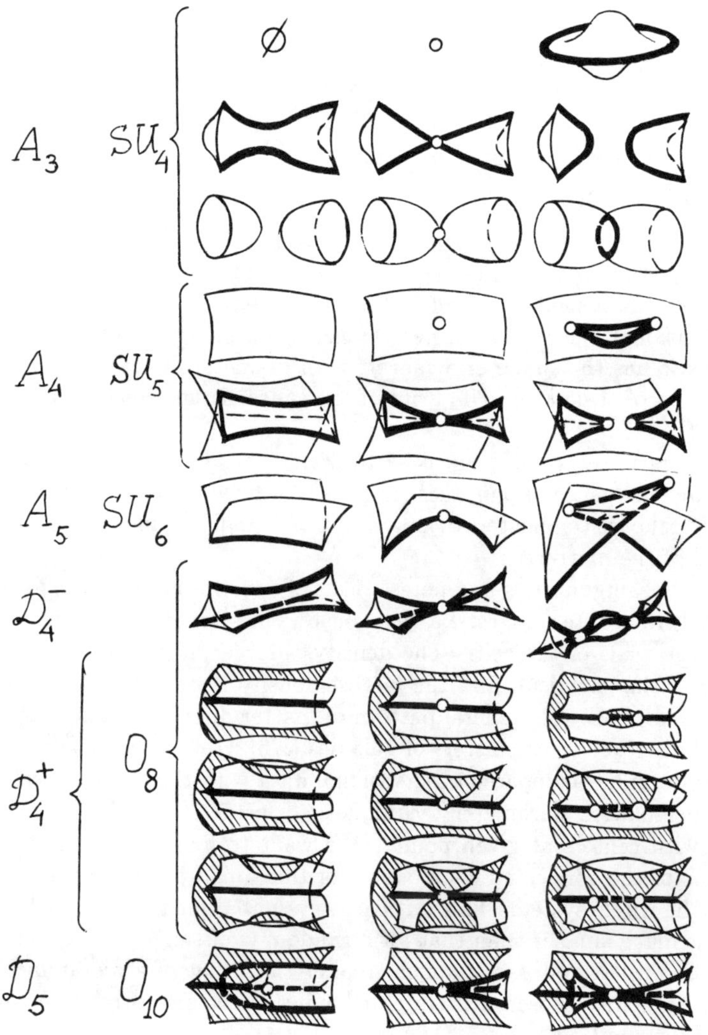

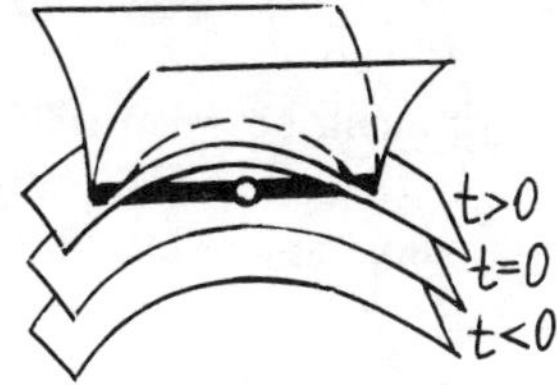

Figure 21: The typical perestroikas of caustics in 3-space

Figure 22: The perestroika of the emergence of a caustic in space–time

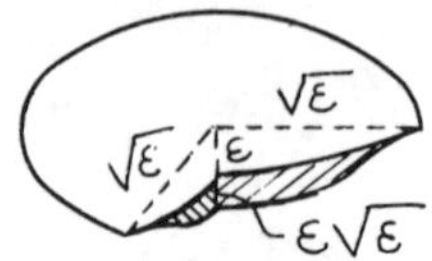

Figure 23: A newborn caustic in 3-space—the saucer or Zeldovich's pancake

moment onwards). At time $t = \epsilon$ this caustic has the shape of a sickle, with length of scale $\sqrt{\epsilon}$. It has two cusps and (generically) two points of inflection. The perestroika of the emergence of a sickle was called 'lips' by R. Thom.

In 4-dimensional space–time the perestroika of the emergence of a caustic is similar: the new caustic has the shape of a saucer, with diameter of scale $\sqrt{\epsilon}$, depth of scale ϵ, and thickness of scale $\epsilon\sqrt{\epsilon}$. The brim of the saucer is an almost elliptical cuspidal edge (Fig. 23).

Such saucers were called 'pancakes' by Zeldovich [69]. He identified the perestroikas of the emergence of pancakes (in potential flows of dustlike matter) with the generation of galaxies (or clusters of galaxies) from small inhomogeneities of the initial velocity field of the matter.

The theory of singularities and perestroikas of caustics predicts the formation of a cellular structure at later stages. Particle clusters are formed on the caustic surfaces which form the walls of the cells. The density is larger at certain lines (the lines of singularities of the caustics), and the largest density is obtained at certain isolated points. It appears that this general pattern of matter distribution is compatible with the observed distribution of clusters of galaxies ([75], [76], [74]).

S.F. Shandarin has compared observational data with the theory of cellular structure, using percolation parameter techniques. Consider the union of all balls with radii r and with centers at given points. We want to find the minimal r for which the diameters of the connected components of this union become comparable to the diameter of the whole system. This critical r, evaluated for the observed distribution of galaxies, is much smaller than that for a random (Poisson) distribution having the same average density, and is also much smaller than that of a hierarchical distribution around randomly chosen centers and having the same correlation functions as the observed distribution.

This shows that the structure of the observed distribution displays a concentration of galaxies along certain surfaces, or at least along certain lines. Thus, a cellular (or, possibly, weblike) structure of the observed galaxies is not an artificial subjective but an objective phenomenon.

In principle it is possible to distinguish between cellular and weblike distributions using another percolation parameter: the maximum value of r for which the union of the balls of radius r separates the whole space into empty regions that are small in comparison to the whole system.

The existing tables of singularities of caustics in higher-dimensional spaces [28] allows one to study events of higher codimension in systems of caustics in physical

2- or 3-space. Under small generic perturbations of the system, such events split into the sequence of standard perestroikas described above. However, if the system depends on parameters, higher order degeneration becomes unavoidable: under a small perturbation of a system depending on parameters, higher order singularities occur, for certain parameter values (close to the initial one), at a certain point of the space (in a neighborhood of the initial point).

2.4 Perestroikas of optical caustics

The verification in laser optics of the general theory of singularities and perestroikas of caustics (J. Nye and J. Hannay [77]) has led to the discovery of certain new and interesting topological properties of optical Lagrangian singularities (Ju.V. Chekanov [78]).

Definition. A Lagrangian submanifold of the space of a Lagrangian fibration is called *optical* if it lies in a hypersurface that transversally intersects the fibers along quadratically convex hypersurfaces in the fibers.

Recall that a fiber of a Lagrangian fibration has a natural affine structure. Hence quadratic convexity (positive definiteness of the second quadratic form) of a hypersurface in a fiber of a Lagrangian fibration is well defined.

Example. The eikonal equation $p^2 = 1$ defines a fiberwise quadratically convex hypersurface in the cotangent bundle of a Riemannian manifold. Hence the solutions of the Hamilton—Jacobi equation $(\nabla u)^2 = 1$ define optical Lagrangian submanifolds.

The optical Lagrangian submanifolds form a very particular class of the Lagrangian submanifolds. However, all stable singularities of Lagrangian maps admit stable optical realisations. Moreover, the generic optical Lagrangian singularities are the same as the generic (unrestricted) Lagrangian singularities (see [79] and the results of I.A. Bogaevski, 1989).

It happens, however, that optical Lagrangian singularities have certain peculiar global properties not present for (unrestricted) Lagrangian singularities.

Example. *The Euler characteristic of a smooth compact set of critical points of the Lagrangian projection of a generic optical Lagrangian submanifold is equal to zero* [78].

Corollary 1. *The components of caustics of optical Lagrangian singularities cannot have the shape of saucers (pancakes).*

Indeed, the corresponding set of critical points is a 2-sphere, and its Euler characteristic equals 2.

Corollary 2. *For optical caustics the emergence of perestroikas of saucer shape is* *impossible.*

Chekanov's proof is based on the following interesting

Lemma. *The characteristics of a fiberwise convex hypersurface containing an optical Lagrangian submanifold are not tangent to the critical set of the Lagrangian projection, at the points of this critical set. For instance, at the smoothness points of the critical set a direction field is defined; it coincides with the field of kernels of the Lagrangian map at the points of types A_μ, $\mu \geq 3$.*

The compete list of *optical perstroikas* of caustics in 3-space contains 7 optical metamorphoses (the above given list of (unrestricted) Lagrangian metamorphoses contains 11 topologically distinct metamorphoses, neglecting the orientation of the time axis).

The 4 optically impossible perestroikas are the types A_3 and D_4^+ (Fig. 21).

In order to describe the topological restrictions on generic optical caustics and perestroikas in 3-space, let us consider the surface of critical points of the Lagrangian projection of an optical Lagrangian 3-manifold. This surface has simple (quadratic) conical singularities at the type D_4 singularities of the Lagrangian map; otherwise (at the type A_μ points) it is smooth.

In the general theory of Lagrangian singularities we distinguish points of 2 types D_4 (two real forms): D_4^+ and D_4^-, according to the sign of the term x_2^3 in the normal form.

In the optical theory we have to distinguish 3 cases. Consider the tangent cone to the surface of critical points at a point of type D_4. It is a nondegenerate quadratic cone in a 3-space (in the tangent space to the kernel of the Lagrangian submanifold). The kernel of the projection to the base space of the Lagrangian fibration at points of type D_4 is a 2-plane. This 2-plane may intersect the cone either along two real lines (the case D_4^+) or at the origin (the lines of intersection are complex conjugate, the case D_4^-). In the first case one real curve of points of type A_3 touches the cone at the origin. In the second case there are three such curves.

The parts of the tangent to this A_3 curve and of the characteristic direction of the ambient fiberwise convex hypersurface at a point of type D_4^+, lying in the same domain bounded by the halfcone and tangent to the critical set, cannot be separated by our 2-plane (that is, by the kernel of the Lagrangian projection). Accordingly, we distinguish the following 2 cases: (D_4^+, drop) and $(D_4^+, \text{triangle})$ (Fig. 24).

The name refers to the behavior of the field of kernel lines of the Lagrangian projection on the surface of critical points. The view of one half of this conical surface seen from the side of the vertex of the cone is presented in Fig. 25. Only the visible part of the kernel direction is presented. The winding of the projection of the field of kernels in the 3 cases is the same as the winding of the tangent lines of either a 'triangle' hypocycloid with 3 cusps (cases (D_4^-) and $(D_4^+, \text{triangle})$) or a 'drop' curve. These 3 types of optical D_4 type singularities correspond to the 3 types of umbilical points on surfaces in euclidean 3-space (Fig. 26)).

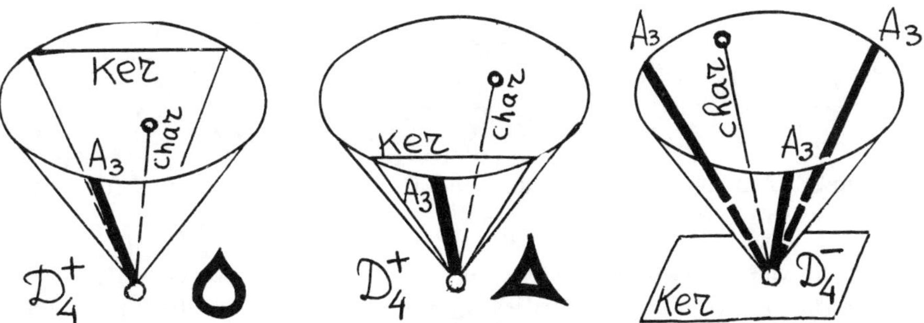

Figure 24: The cone A_2 of critical points, the kernel Ker of the Lagrangian projection, and a characteristic of the ambient hypersurface, for the 3 variants of type D_4 singularities

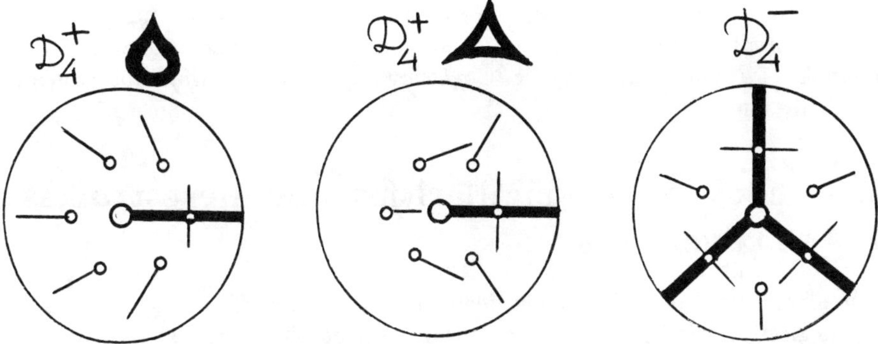

Figure 25: The field of kernels of the projection on the cone of critical points

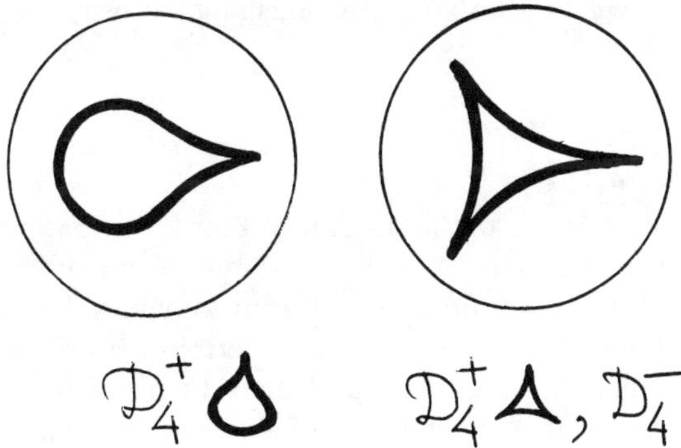

Figure 26: The triangle and the drop

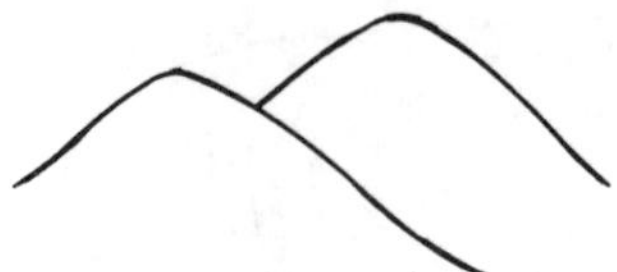

Figure 27: A nonsmooth maximum function of a smooth family of functions

The Euler characteristic χ of a compact critical surface of an optical Lagrangian projection (having only the singularities A_μ and D_4) is given by Chekanov's formula:

$$\chi + 2(\#(D_4^-) + \#(D_4^+, \text{triangle})) = 0.$$

Moreover, the lefthand side (whose value can be nonzero) is preserved under optical perestroikas of critical surfaces with boundary, provided that the boundary is not affected by the perestroika and that all critical points are of type A_μ or D_4.

The impossibility of the 'emergence of a flying saucer' under perestroika of an optical caustic in 3-space implies the impossibility of the 'emergence of a lips' under perestroika of a plane optical caustic. Chekanov has also proved that 'lips' components are globally impossible.

2.5   Shock wave singularities and perestroikas of Maxwell sets

The noninteracting particles model, leading to clustering of caustics of Lagrangian maps, neglects effects of collision or scattering of nearby particles. For instance, in the 1-dimensional case, the 'particles' are supposed to go trough each other.

Models taking into account (inelastic) collisions lead to different scenarios—to the formation of shock waves at spots of collision. One of the simplest models of this kind is described by the *Burgers equation* with vanishing viscosity:

$$\frac{\partial u}{\partial t} + u\frac{\partial u}{\partial x} = \epsilon\Delta u, \qquad \epsilon \to 0,$$

for a potential vector field $u = \nabla S$.

The well-known description of the shock waves for this equation relates these waves to singularities of the maximum function of a family of smooth functions.

The dependence on the parameters of the maximum value of the function of a family is continuous, but can be nonsmooth for certain parameter values; namely, for those for which this maximum is attained at more than one point.

Consider, for instance, the horizon line of a landscape. This line has a break at the point of intersection of the visible contours of two hills (Fig. 27).

The parameter values for which the maximum function is not smooth on the parameter space form, generically, a hypersurface in the parameter space. We will call this hypersurface the (small) *Maxwell set* of the family (because of Maxwell's

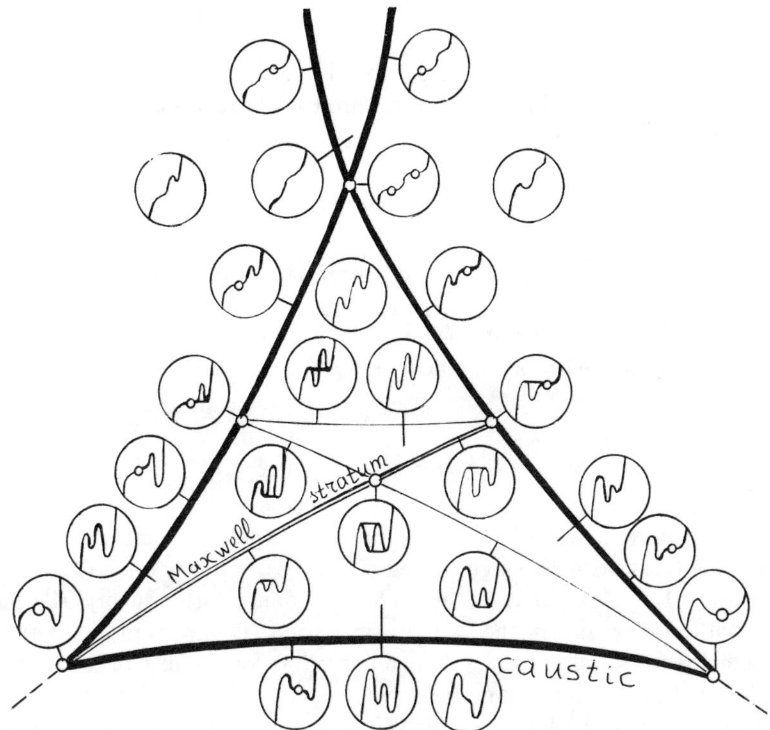

Figure 28: The caustic and Maxwell strata of the family $x^5 - x^3 + bx^2 + cx$

rule in the Van der Waals theory, according to which phase transition occurs at a parameter value for which two maxima of a certain smooth function are of equal height). This small Maxwell set is part of the large Maxwell set $2A_1$, consisting of the points in the parameter space for which the function has two equal critical values. Both Maxwell sets live in the same space as the caustic defined by the family (the caustic being the set A_2 of those points in the parameter space for which the function has a nonMorse critical point).

Example. The caustic and the (local) Maxwell set of the family $x^5 + ax^3 + bx^2 + cx$ are diffeomorphic surfaces in the parameter space (a, b, c) if we take into account the complex critical points. In Fig. 28 sections of these surfaces with $a = -1$ are given. The small Maxwell set is represented by the double line, while the complex Maxwell set is represented by the dotted line.

Both the caustic surface and the analytic continuation of the Maxwell set of this family are diffeomorphic to the swallowtail surface.

Remark. The (large) Maxwell surface divides the pyramid formed by the polynomials in our example with only real critical points into 5 parts (whose intersections with the plane $a = -1$ are visible in Fig. 28).

The 'updown' sequence ($k_\mu = 1, 1, 1, 2, 5, 16, 61, 272, \ldots$ for $\mu = 0, 1, 2, \ldots$) of numbers of parts of the pyramid for general A_μ families of functions ($f = x^{\mu+1} + q_1 x^{\mu-1} + \cdots + q_{\mu-1}x$) generates the function

$$\sum_{\mu=0}^{\infty} k(\mu)\frac{t^\mu}{\mu!} = \tan t + \sec t.$$

The number of components of the complement of the union of the caustic and the (large) Maxwell set is equal to $k_\mu + k_{\mu-2} + k_{\mu-4} + \cdots$ [80].

These results provide natural generalisations of the Bernoulli and Euler numbers: replace A_μ by any sequence of singularities, for instance D_μ.

Excercise. Prove that
$$k(B_\mu) = k(C_\mu) = k(A_{\mu+1}).$$

The small Maxwell sets of generic families of functions depending on 2 parameters are planar curves with singularities of 2 types: endpoints (A_3) and Y-shaped triple points ($3A_1$). Here, as in other problems of singularity theory, the number of arguments of the functions in the family is irrelevant (it may even be infinite, i.e. the statement holds for families of functionals as well as for families of functions). What is important is the number of parameters.

Suppose now that the family of functions is time dependent. A time-dependent function (depending on k parameters) may be regarded as a function depending on $k + 1$ parameters. The Maxwell set of a time-dependent family of functions (of k parameters) may thus be regarded as a hypersurface in the $(k + 1)$-dimensional 'parameter space–time', while the momentary Maxwell sets are the sections of this hypersurface by isochrones $t = $ const.

Hence, in order to study the generic perestroikas of the momentary Maxwell sets in a k-space, we have to investigate the generic perestroikas of the sections of the generic Maxwell sets in a $(k + 1)$-space.

This study was undertaken by Bogaevski for the cases $k = 2$ (perestroikas of lines in a parameter plane) and $k = 3$ ([80]). The answer for $k = 2$ is presented in Fig. 29 ([81], [82]).

There are 5 topologically distinct germs of generic Maxwell surfaces in 3-space. The generic perestroikas of the sections of these surfaces by isochrones depend on the position of the critical isochrone surface with respect to the Maxwell set. The 10 generic perestroikas are shown in Fig. 29 (if the cases differing only by the direction of time are counted separately, the number of distinct perestroikas becomes 18).

We now return to shock waves. The *shock waves* for the Burgers equation with vanishing viscosity are the Maxwell sets for a specific family of functions; namely,

$$F(x, t) = \min_y f(y; t, x), \qquad f = \frac{(x - y)^2}{2t} - S_0(y)$$

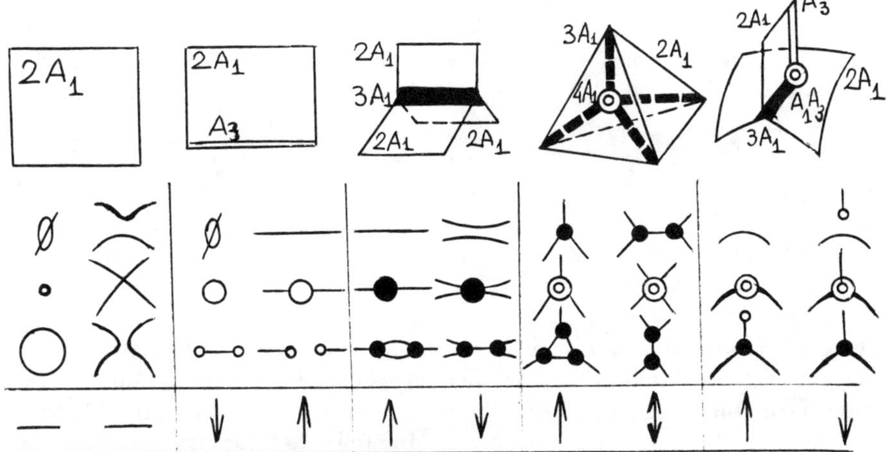

Figure 29: The typical singularities of Maxwell surfaces in 3-space, and the typical perestroikas of Maxwell curves and shock waves in the plane

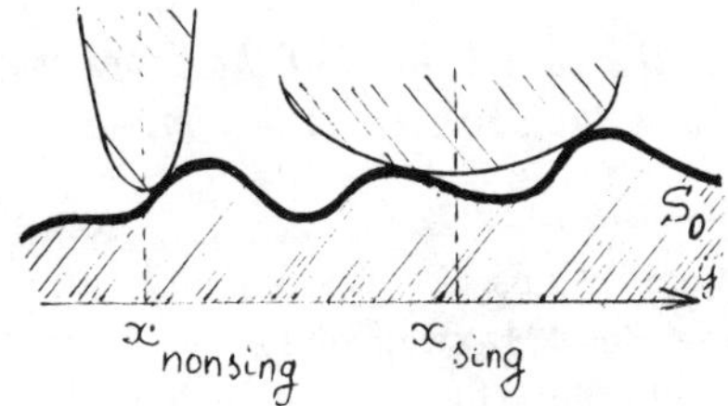

Figure 30: The singular and nonsingular positions x of the axis of a paraboloid

(this follows from the so-called Coula—Hopf transformation, reducing the Burgers equation to the heat equation and described in Forsyth's textbook [83] many years before its rediscovery by Florin [84]). In the space–time $\{(t,x)\}$ the shock waves sweep the Maxwell set of the above given family of functions f of the physical point y and parameters (t,x). The momentary shock waves are the intersections of this Maxwell set with the isochrones.

The minimum function F may be geometrically described as follows. Push downwards a paraboloid with vertical axis over a point x and with width depending on the time t until it touches the graph of the initial potential function S_0. The height of the paraboloid at which it first touches this graph is the required function of x.

If t is very small, the paraboloid $z = (x-y)^2/2t$ is very thin. In this case the point of contact between the paraboloid and the graph is unique. It depends smoothly on x, hence F is a smooth function for t small. For t larger the paraboloid widens. In this case, for certain x there will be more than one point of contact (Fig. 30). These special x's form the momentary shock wave (and the momentary Maxwell set of our family).

Experiments with these paraboloids show that for generic initial surfaces (potentials S_0) only the generic singularities and perestroikas in Bogaevski's list occur. However, as Gurbatov and Saichev [85] have observed, the set of generic perestroikas of shock waves is smaller than that in Bogaevski's list of generic perestroikas of Maxwell sets.

Certain perestroikas of Maxwell sets occur as perestroikas of shock waves only in one direction of time evolution (for instance, a triangle may collapse but cannot emerge); other perestroikas of Maxwell sets occur in both directions of time evolution; and still others do not occur at all. These admissible directions of perestroikas are indicated by arrows in Fig. 29.

These admissible directions depend on convexity properties of the restriction to the impulse space of the Hamilton function defining the corresponding Hamilton—Jacobi equation. (For Burgers equation this Hamilton function is $p^2/2$.) Namely, necessary and sufficient conditions for a perestroika of Maxwell sets to occur as a perestroika of shock waves (in a system defined by a fiberwise convex Hamilton function) are:

1. *A local shock wave born at the moment of perestroika is contractible in a neighborhood of the point of perestroika at the next moment* (Bogaevski's theorem).

2. *The homotopy type of the complement of the shock wave immediately after the moment of perestroika is the same as at this moment* (Ju. Baryshnikov's theorem).

Every one of these two conditions is necessary and sufficient for the admissibility of a perestroika of time-dependent Maxwell sets on a plane or in 3-space (Bogaevski), However, it is not known whether this is true in a higher-dimensional space, and whether the above conditions are equivalent for higher-dimensional shock waves governed by Burgers's equation with vanishing viscosity.

Remark. Both Chekanov's (§ 2.4) and Bogaevski's (§ 2.5) theory suggest that the embedding of a Lagrangian manifold in a fiberwise convex hypersurface implies severe topological restrictions. Thus it seems that the optical topology is rather different from the general symplectic topology (see also [86]–[89]).

Problem. Jacobi [90] has mentioned that any caustic of the family of geodesics starting at a generic point of an ellipsoid has at least 4 cusps. Is this true for other riemannian metrics on spheres (for instance, for generic metrics close to the standard metric)? This 4-cusp property, if true, should be a generalisation to symplectic topology of the 4-vertices theorem, according to which a closed planar curve has at least 4 points of stationary curvature (see [91], [92]).

Chapter 3

Contact geometry

Contact geometry forms the mathematical basis of geometrical optics, in the same sense in which symplectic geometry forms the basis of classical mechanics. The 'optical–mechanical analogy' of Hamilton allows one to translate problems and results from symplectic geometry to the language of contact geometry and vice versa. However, a direct approach in terms of contact geometry is in many cases preferable, at least from the point of view of geometrical intuition: it shows the geometrical content of formulas from the symplectic theory. The relation between symplectic and contact geometry is similar to the relation between the geometries of linear spaces and projective geometry: in order to obtain a contact counterpart of a symplectic piece of theory one has to replace functions by hypersurfaces, affine spaces by projective spaces, etc.

In this chapter contact geometry is used for the study of singularities and perestroikas of wave fronts.

3.1 Wave fronts

Caustics can be described as traces sweeped by cuspidal edges of moving wave fronts. The theory of singularities of wave fronts is a particular case of the general theory of Legendre singularities in contact geometry. This general theory implies a classification of the singularities of Legendre transformations of smooth functions and of dual hypersurfaces of smooth projective surfaces.

Definition. A *contact structure* on a manifold is a field of tangent hyperplanes (*contact hyperplanes*) that is nondegenerate at any point (Fig. 31).

Locally such a field is defined as the field of zeros of a 1-form α, called a *contact form*. The nondegeneracy condition is:

$d\alpha$ is nondegenerate on the hyperplanes on which α vanishes; equivalently, in $(2n + 1)$-space:

$$\alpha \wedge (d\alpha)^n \neq 0.$$

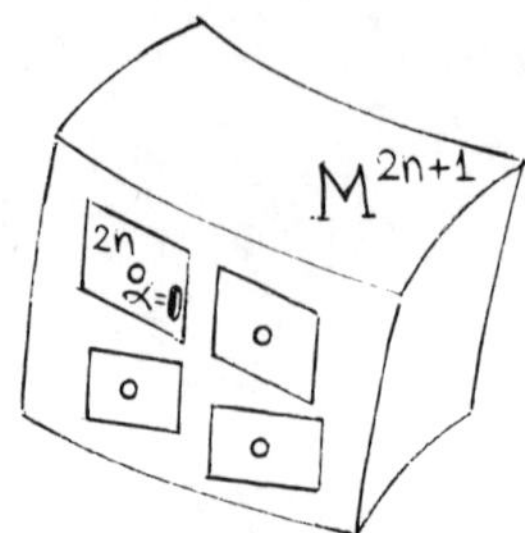

Figure 31: A contact structure: a maximally nonintegrable field of tangent hyper-planes

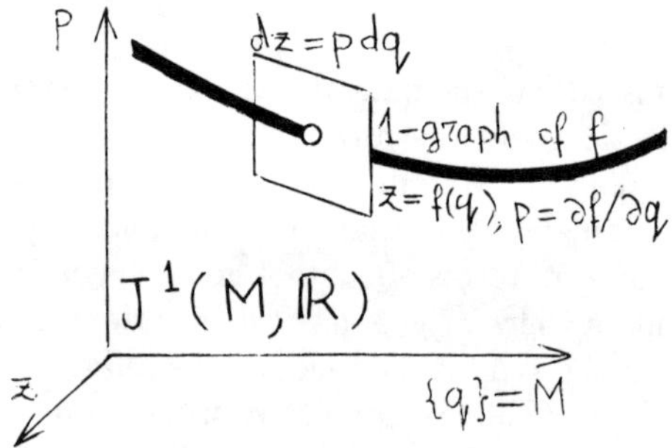

Figure 32: The contact structure of the manifold of 1-jets of functions

This condition can be regarded as a condition for maximal nonintegrability, since Frobenius's integrability condition for the field of hyperplanes $\alpha = 0$ is $\alpha \wedge d\alpha = 0$.

Example 1. $\mathbf{R}^{2n+1}$, $\alpha = dz - p_1\, dq_1 - \cdots - p_n\, dq_n$. *All contact structures on a manifold of fixed dimension are locally contactomorphic, and all contact forms are locally diffeomorphic* (and hence may be written in the above normal form, using suitable *Darboux local coordinates*; see, for example, [93], [14]).

Example 2. The *manifold* $J^1(M, \mathbf{R})$ *of 1-jets* of functions $M \to \mathbf{R}$ (Fig. 32) has a natural contact structure, '$dz - p\, dq$' (a 1-jet (z, p, q) of a function f is its Taylor polynomial $f(x) = z + p(x - q) + \cdots$ of degree 1).

Example 3. A *contact element* on V is a hyperplane in the tangent space of V. All contact elements on V form a bundle over V, with as fiber the projective space of contact elements centered at the same point of contact. The *bundle of contact elements* on V is the projectivisation of the cotangent bundle of V. Thus, the total space of the *projectivised cotangent bundle PT^*V* has a natural contact structure (Fig. 33).

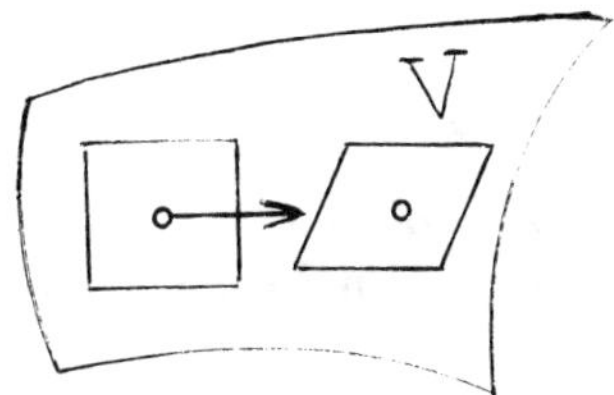

Figure 33: The contact structure f the manifold of contact elements on V

The *natural contact structures* in examples 2 and 3 are defined by 'integrability conditions'.

The velocity of motion of a contact element belongs to the contact hyperplane if and only if the velocity of the motion of the contact point belongs to the contact element.

The contact structure of the space of 1-jets of functions is defined by the following condition: the manifold of 1-jets of a function at all points of the underlying manifold is tangent to the hyperplane at any point.

This natural contact structure of the space of 1-jets of functions can in local coordinates (z, p, q) be described as the field of zeros of the contact 1-form $dz - p\,dq$ (here z denotes the value of the function, q is a point of the underlying manifold, and p is the value of the first differential of the function at q).

Givental's theorem holds for submanifolds of a contact space: *the local exterior geometry is determined by the local interior geometry* ([13], [14]).

On a submanifold of a contact manifold the contact structure defines a Pfaffian structure. A *Pfaffian equation* is an equation of the form $\alpha = 0$, where α is a differential 1-form. A *Pfaffian structure* is (locally) a class of equivalent Pfaffian equations (two Pfaffian equations are called *equivalent* if the corresponding 1-forms only differ by multiplication with a nowhere vanishing function). A Pfaffian structure defines a field of tangent hyperplanes at the points where the Pfaffian forms do not vanish. In a neighborhood of such a nondegenerate point the Pfaffian structure is, in essence, the same as its corresponding field of tangent hyperplanes.

Example. The Pfaffian structures defined by the forms $x\,dx$ and $x^2\,dx$ coincide in a neighborhood of any nonzero point, but are different at $x = 0$.

Theorem 1. *The germ of a submanifold of a contact space is determined, up to a contactomorphism, by the Pfaffian structure induced on the submanifold by the contact structure of the ambient space.*

Theorem 2. *Suppose that the germs of the restrictions to a submanifold of two contact structures in space coincide. Then there exists a local diffeomorphism of the space mapping the first form to the second and fixing the points of the submanifold.*

For a proof see [14].

Definition. A *Legendre submanifold* of a contact manifold is an integral submanifold of maximal dimension (that is, equal to n for a $(2n + 1)$-dimensional manifold).

Example 1. The fibers of the fibration $PT^*V \twoheadrightarrow V$ are Legendre submanifolds.

Example 2. The set of 1-jets of a function on M is a Legendre submanifold of the manifold $J^1(M, \mathbf{R})$ of 1-jets of functions on M.

Example 3. The set of contact elements of V tangent to a fixed smooth submanifold (of arbitrary dimension) in V is a Legendre submanifold of the space of contact elements on V.

Definition. A *Legendre fibration* is a fibration with Legendre fibers.

Example 1. The projective cotangent fibration $PT^*V \twoheadrightarrow V$ associating to a contact element its point of contact is a Legendre fibration.

Example 2. The fibration $J^1(M, \mathbf{R}) \twoheadrightarrow J^0(M, \mathbf{R})$ of the space of 1-jets of functions on M over the space of 0-jets, defined by the 'map of forgetting derivatives', $(q, p, z) \mapsto (q, z)$, is a Legendre fibration.

Theorem. *All Legendre fibrations of the same dimension are locally Legendre equivalent (in a neighborhood of a point of the total space).*

We project a hyperplane defining a contact structure from a point of the total space of a Legendre fibration along the fibers into the base space. The projection is a contact element tangent to the base space at the point which is the projection of the fiber.

Thus we have constructed a map from the total space of a Legendre fibration to the space of contact elements of the base space. This map is a (local) diffeomorphism, since nondegeneracy of the contact structure implies that the projection of the hyperplane defining the contact structure turns with nonzero velocity if the point of the total space moves with nonzero velocity along the fiber.

This map transforms the initial contact structure and Legendre fibration to the natural Legendre fibration of the space of contact elements of the base, proving the theorem.

Remark. The fibers of a Legendre fibration are locally equipped with a natural projective structure (defined in the proof given above). This structure is an analog of the natural affine structure in the fibers of a Lagrangian fibration in symplectic geometry.

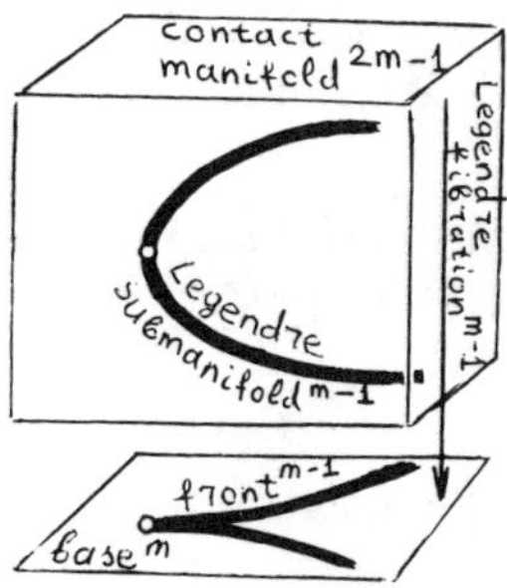

Figure 34: A Legendre map and its front

The projective structure of the fibers of a Legendre fibration even has a more geometrical content, since every map of Legendre fibrations (preserving the contact structure and mapping fibers to fibers) automatically induces a unique projective map of the fibers (defined by the action of the diffeomorphism of the base space on the contact elements of the base). In the symplectic case affine maps of fibers are defined only up to translations.

Definition. The projection of a Legendre submanifold of the total space of a Legendre fibration to the base space of this fibration is called a *Legendre map* (Fig. 34). The image of a Legendre map is called the *front* of this map.

Generically, a front is a hypersurface of the base space of the Legendre fibration. The exceptional Legendre submanifolds (for instance, the fibers, whose fronts are the points of the base space) form a set of infinite codimension in the space of all Legendre manifolds of the total space of the fibration.

Example 1 (*Legendre duality*). Consider the space of contact elements of the projective space P^n. This space is naturally isomorphic to the space of contact elements of the dual projective space P^{n*}:

$$PT^*(P^n) \approx PT^*(P^{n*}). \tag{1}$$

Indeed, a point of the first space can be considered as a pair formed by a point of P^n and a hyperplane containing this point. The isomorphism sends it to the pair formed by this hyperplane and this point.

The geometrical base of the theory of Legendre transformations is the following elementary

Theorem. *The natural isomorphism (1) sends the natural contact structure of the first space to that of the second.*

Proof. Consider a smooth submanifold of the product of a manifold with itself, having contact with the diagonal. We will use the evident

Lemma. *The tangent space of the above chosen submanifold at a point of the diagonal is invariant under the involution of the product given by permutation of factors.*

Consider the flag manifold PT^*P^n (point $\subset$ hyperplane) in P^n. A flag is said to be *dominated* by another flag if the point of the latter belongs to the hyperplane of the first. The graph of this relation is a smooth submanifold of the product of the flag manifold with itself, and contains the diagonal. Permutation of factors transforms the graph of the relation 'to dominate' to the graph of the relation 'to be dominated'.

The tangent planes of both graphs at a point of the diagonal coincide, by the lemma. Hence the intersections of these tangent planes with the tangent plane of the product of the first factor with this point (at this point on the diagonal) coincide. But these two intersections define the natural contact structures of the first factor, considered as $PT^*(P^n)$ for one of the intersections, and considered as $PT^*(P^{n*})$ for the second one. Coincidence of the intersections now gives the theorem.

Remark. A similar reasoning proves a similar theorem for complete flags (consisting of the projective spaces of arbitrary dimensions). This duality theorem is the base of the projective duality of curves in a projective space P^n (the dual curve lies in P^{n*} and is the cuspidal edge of the hypersurface formed by the tangent planes to the initial curve).

Returning to Legendre duality, let us consider the two natural fibrations of the flag space (point, hyperplane):

$$
\begin{array}{ccc}
PT^*(P^n) & \longrightarrow\!\!\!\!\!\to & P^n \\
\updownarrow & & \\
PT^*(P^{n*}) & \longrightarrow\!\!\!\!\!\to_\pi & P^{n*}
\end{array}
$$

According to the theorem, both fibrations are Legendre fibrations.

The set of all contact elements tangent to a given hypersurface $H^{n-1} \subset P^n$ is a Legendre submanifold of the flag space. Hence the restriction of π to this Legendre submanifold is a Legendre map (Fig. 35).

The front H^* of this map is called the *hypersurface dual* to H. It lives in the dual projective space: $H^* \subset P^{n*}$. The affine version of the construction of the front of a given hypersurface is called the *Legendre transformation*.

The above theorem implies that the construction of the dual hypersurface (and hence of the Legendre transformation) is involutive:

Corollary. $H^{**} = H$.

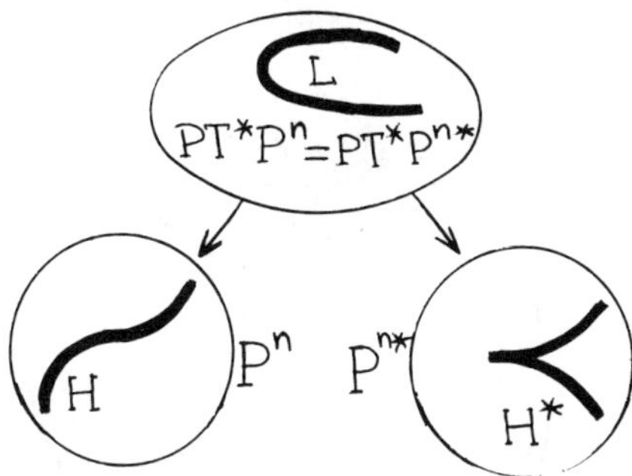

Figure 35: The projective duality between a hypersurface (H) and its front (H^*)

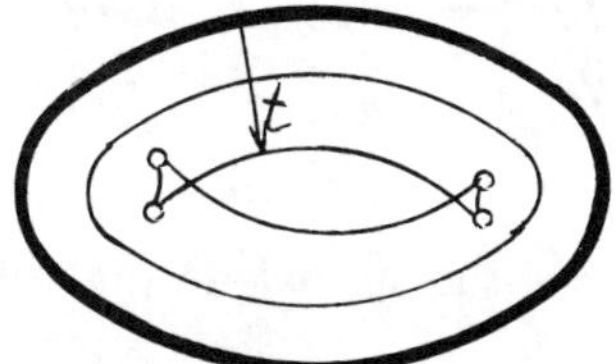

Figure 36: The equidistant curves of an ellipse

Instead of a hypersurface H we may begin with a submanifold of arbitrary dimension in a projective space: it always defines a Legendre map. For instance, the front of a curve in a projective space (that is, the front of the corresponding Legendre map) is the manifold of its tangent hyperplanes.

Example 2 (*equidistants*). For a smooth hypersurface H in a euclidean space $\mathbf{R}^n$ we fix, for each normal, a point at distance t. This gives a Legendre map $H \hookrightarrow PT^*\mathbf{R}^n \twoheadrightarrow \mathbf{R}^n$. Its front is called the *equidistant hypersurface* of the given hypersurface H (Fig. 36).

Locally, every Legendre map is equivalent to a map as in example 1, and also to a map as in example 2. The front (of the Legendre map of a submanifold of $J^1(M, \mathbf{R})$ or PT^*V) uniquely determines the Legendre map (with the possible exception of a set of Legendre maps of infinite codimension in the space of all Legendre maps). Hence the (local) theory of Legendre singularities coincides with the theory of singularities of Legendre transformations, and also with the theory of singularities of equidistant hypersurfaces.

The general theory of Legendre singularities also allows one to transform any result in the geometry of wave fronts (i.e. equidistant hypersurfaces) to a result in the geometry of Legendre transformations, and vice versa. Clearly, one may also apply this to cases in which the contact structure is less obvious.

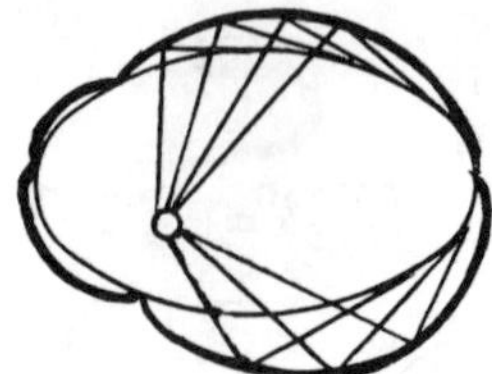

Figure 37: The pedal curve (podhere) of an ellipse

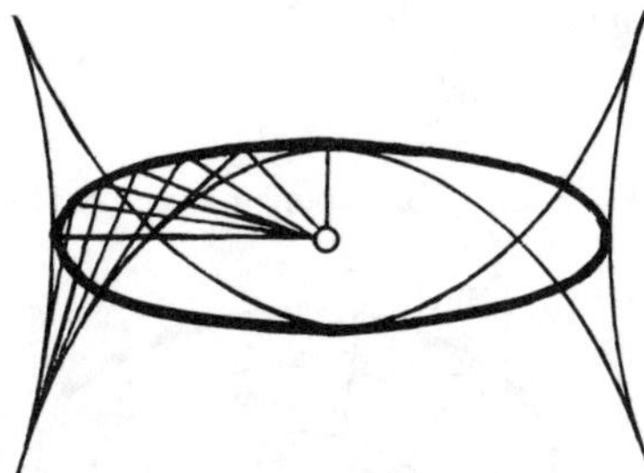

Figure 38: The envelope of the normals to the radius vectors of an ellipse

Example 3 (*pedal hypersurfaces*). Consider a smooth hypersurface in a euclidean space, not containing a point O. The *pedal hypersurface* is the set of bases of the normals drawn from O to the tangent hyperplanes of the hypersurface (Fig. 37).

A pedal hypersurface may have singularities.

Problem. Prove that the singularities of the pedal hypersurfaces of generic hypersurfaces coincide (up to a diffeomorphism) with the singularities of the fronts of generic Legendre maps (and hence with the singularities of the dual hypersurfaces of generic smooth hypersurfaces, and with the singularities of the graphs of Legendre transformations of generic smooth functions).

Hint. The pedal surface is an inversion of the front of the given hypersurface, i.e. it may be obtained from the dual hypersurface by an inversion. An inversion is a diffeomorphism outside the origin.

Example 4 (*the envelope of normal hyperplanes*). Consider the hyperplanes normal to the radius vectors of a hypersurface (drawn from a point O outside the hypersurface to the points of the hypersurface (Fig. 38)). The envelope of this family may have singularities (these were studied by A. Cayley in the case of planar ellipses or spatial ellipsoids).

Problem. Prove that the singularities of the envelope of normal hyperplanes, constructed from a generic hypersurface, are diffeomorphic to the singularities of the where (p, q) are the usual coordinates in the cotangent bundle and where the Lagrangian projection sends (p, q) to q. For instance, we may choose $F = \pm x^4 + q_1 x^2 + q_2 x$.

The functions of this family corresponding to points q that are not the cusp do not have singularities more complicated than A_2. Hence a loop bounding a disk in the q-plane with center at a cusp has an index. This index is 1 for a cusp of one type, and -1 for the cusp of the other type.

To be more precise, we have to choose the natural orientation of the q-plane in order to define the loop.

In the definition of the index we have used the condition that the function 'behaves at infinity like x'; if it would behave at infinity like $-x$, the index would become of opposite sign. Hence we choose the positive direction of the x-axis of the generating family in such a way that $F_x(x, 0)$ is nonnegative. This choice determines an orientation of the kernel of the projection, and hence of the fold line. Thus, at a Lagrangian cuspidal point, the cuspidal line of critical values has a natural orientation. This orientation determines the natural orientation of the base plane on which the cuspidal line lives: the cuspidal line should be oriented as the boundary of the horn-like domain bounded by it.

The boundary index of a generic 2-parameter family of functions of one variable gives a lower bound for the number of cusps of the projection of the Lagrangian surface defined by this generating family of functions.

5.5 Global properties of singularities

Let us start with an example: consider the Gauss map of a cooriented surface in euclidean 3-space. This Lagrangian map sends a point of the surface to the unit normal vector based at the origin. If the surface is quadratically convex, the Gauss map has no singularities. The Jacobian function is the Gaussian curvature of the surface. The *Minkowski problem* is the problem of determining a convex surface whose Gaussian curvature is known as a function on the Gauss sphere (the target sphere).

It can be easily seen that this function should be positive and that the center of mass of the sphere should lie at the origin (if the mass distribution is given by this function). It is a deep theorem (of Minkowski, A.D. Aleksandrov, Pogorelov,...) that a convex surface can be reconstructed from the given Gaussian curvature on the Gauss sphere.

If the surface is not convex (it may even have a nonspherical topology), the restriction on the center of mass still applies. The formulation of Minkowski's problem requires a more refined description of the data, however. I suggest as data for the nonconvex Minkowski problem: a smooth map

$$f : (M, \tau^n) \to S^n,$$

where S^n is the standard Gauss sphere and M^n is a closed manifold without boundary, equipped with a nondegenerate volume element (n-form) τ. The direct image $f_!\tau$ is a density on the Gauss sphere whose center of mass should lie at the origin. In the convex case f is the identity map and $\tau = \omega/K$, where ω is the standard area element and K the Gaussian curvature of the sphere.

Thus, in the general case, on M the Gaussian curvature is replaced by the Jacobian function $J = (f^*\omega)/\tau$ of f. Hence the following problem arises: *which functions on M are the Jacobian of a smooth map from (M, τ) to the standard sphere?* Of course, the integral of such a function should be equal to the volume of the standard sphere. Are there other restrictions? Of course, *similar problems arise for Lagrangian maps (the Gauss map is a particular case) and for Gauss maps (of immersed or embedded hypersurfaces).*

These problems are far from being solved. Let us consider a most simple version: which functions can be the Jacobian of a map from a sphere to the plane? In other words, which differential 2-forms on the sphere can be represented as the product of two closed 1-forms? Locally such a representation is always possible, but for the global representability the 2-form should be exact (its integral should vanish).

Theorem. *Any generic exact 2-form on a 2-sphere is induced by the area element of the euclidean plane under a generic smooth map from the sphere to the plane.*

(Compare with [139].)

Consider the zeros of a given generic exact 2-form. The first genericity condition requires that these make up a smooth curve (the form being $h\tau$ with τ the area element, the requirement is that 0 is not a critical value of the smooth function h). The components of this curve split the sphere into regions on which the Jacobian is sign-definite. The second genericity condition requires that the integrals of the form along these components should be nonresonant (i.e. they should not satisfy a finite set of linear homogeneous equations with integer coefficients). For instance, rational independence of the integrals suffices.

Genericity of the map means that its only singularities are Whitney folds and cusps (generically located; this implies that the map is stable).

Definition. The *tree of a generic form* is a graph whose vertices are the components on which the Jacobian is sign-definite, and whose edges join those components that are separated by a component of the curve of zeros.

Example. The tree of the standard projection of the sphere to its equatorial plane is $\bullet - \bullet$, the vertices being the northern and southern hemispheres and the edge being the equator.

A generic form on a sphere defines a tree and a positive function on this tree (i.e. on the set of vertices): the integrals along the components, oriented by the form. The alternating sum of the values of this function should vanish. The tree and this function are the only invariants of a generic 2-form on a 2-sphere.

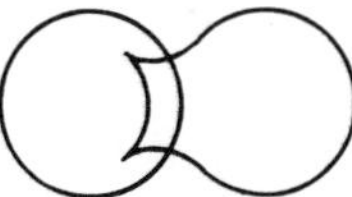

Figure 71: Realisation of a tree with 3 vertices by a map from the sphere into the plane

Definition. The *defect of a function* on a rooted tree is the difference between the sum of its values at the even vertices (joined to the root by an even number of edges) and the sum of the values at the odd (i.e. other) vertices.

Any positive function with defect 0 on some tree admits a realisation by a nondegenerate 2-form on a 2-sphere.

Definition. The *defect of a rooted tree* is the defect of the constant function 1 (i.e. the difference between the numbers of even and odd vertices).

Theorem. *A generic 2-form on a 2-sphere is induced by the area element in the plane under a map whose only singularities are folds if and only if its tree has defect 0.*

The number of cusps of a generic map inducing a form with a given tree is bounded below by twice the defect of the tree.

In fact, there always exists a way of distributing the signs of the cusps such that the number of cusps counted with sign is equal to twice the defect of the tree [140].

Namely, the cusp is *positive* if the component that is diffeomorphically sent to the inside of the horn formed by the line of critical values in a neighborhood of the cusp is even.

Example. The tree $\circ - \bullet - \bullet$, whose defect is 1, admits a realisation by the map (from a 2-sphere into the plane) shown in Fig. 71. Both cusps are positive.

Similar results hold for maps from other surfaces to the plane, provided the components of sign-definitiveness are punctured disks. The graph corresponding to such a form on an orientable surface of genus g is homotopy equivalent to a bouquet of g circles. It has no odd cycles, hence its defect is well-defined.

5.6 Topology of Lagrangian inclusions

Consider a map from a smooth manifold to a symplectic space. Such a map is called *isotropic* if it induces a 0-form from the symplectic structure.

Example. An immersion of a Lagrangian manifold (or of a submanifold of it) is an isotropic map.

A natural way to define Lagrangian varieties is to consider isotropic maps of manifolds having the appropriate dimension (equal to one-half of the dimension of the ambient symplectic manifold). Such a map is called a *Lagrangian inclusion* if its singular points form a subvariety of smaller dimension.

Example. Consider the space of the conormal bundle of the semicubic parabola:

$$\{(p,q) : q_1 = t^2, \quad q_2 = t^3, \quad p_1 2 + p_2 3 t = 0\}.$$

It can be parametrised by the formulas

$$q_1 = t^2, \qquad q_2 = t^3, \qquad p_1 = -3st, \qquad p_2 = 2s,$$

which defines an isotropic map. Its image is a surface in the symplectic 4-space whose only singularity is the origin. It is homeomorphic to the 2-plane.

Givental [141] has called this surface the *open umbrella*. Its projection to the 3-space along the p_2-axis is the usual *Whitney—Cayley umbrella*.

The open umbrella is a Lagrangian surface with one singular point. This point can be eliminated by smoothing in the class of all surfaces in 4-space (since the intersection of the surface with a small 3-sphere centered at the singular point is unknotted). However, it cannot be smoothened in the class of Lagrangian subvarieties, since the Maslov index of a loop bounding a neighborhood of the singular point does not vanish (it is equal to ± 2, depending on the orientation).

Givental has proved that this singularity of isotropic inclusion is stable: any neighboring isotropic inclusion can be reduced to the form of the above example in a neighboring point of the initial surface by a diffeomorphism of the surface and a symplectomorphism of the space.

Givental's conjecture. *The Lagrangian inclusions of surfaces whose only singularities are open umbrellas (and, of course, selfintersections) are dense in the space of all Lagrangian inclusions.*

On the level of jets, the bad set in this problem has codimension 7, hence it seems that a generic Lagrangian inclusion of dimension 2 will not intersect it. However, the proof cannot be reduced to the usual transversality arguments, since the equation defining the isotropic embedding ($f^*\omega = 0$, where f is the inclusion and ω the symplectic form) is *quadratic* with respect to f.

Thus, the conjecture deals with 'generic' maps into a variety defined by a system of quadratic homogeneous equations. This variety is a very bad cone, which may have components of various dimensions, and there are many, a priori inequivalent, natural definitions of 'genericity' of maps into this cone (the same difficulty occurs in the problem of classifying the 'generic' Lie algebras of a given dimension, in which the Jacobi identity is a system of quadratic equations).

Perhaps the simplest problem of this kind is the problem of classifying the 'generic' maps from a plane to a plane that have identically vanishing Jacobians.

The open umbrella appears in the theory of ray systems in the following situation. Consider a hypersurface in a $2n$-dimensional symplectic space and an $(n-1)$-dimensional isotropic submanifold in this hypersurface (we will call it the *initial manifold*).

The rays (the characteristics of the hypersurface) issuing from the points of the initial manifold (locally) form a subvariety of the $(2n-2)$-dimensional symplectic manifold of characteristics. It is isotropic and has dimension $n-1$, generically. If $n = 3$, this submanifold is a Lagrangian inclusion of a surface. Givental has proved [8] that the only singularities of the corresponding Lagrangian inclusions are open umbrellas (provided that the initial manifold belongs to an open dense set in the space of all submanifolds of dimension $n-1$).

The open umbrellas of higher dimension are the spaces of the conormal bundles of open, or 'unfurled', swallowtails of higher dimension. The *unfurled swallowtail* of dimension m is the set of polynomials

$$\{x^{2m+1} + a_1 x^{2m-1} + \cdots + a_{2m} = (x - t)^{m+1}(x^m + \cdots)\}$$

having a root of multiplicity larger than m. It is a Lagrangian manifold in this space of odd-degree polynomials, equipped with the symplectic structure introduced in § 1.1.

Example. The 1-dimensional unfurled swallowtail is the semicubic parabola in the plane. The 2-dimensional unfurled swallowtail in the 4-space of polynomials of degree 5 is shown in Fig. 10.

The natural projection (defined by multiple differentiation of polynomials) sends the $2m$-dimensional space of polynomials of degree $2m+1$ to the $(m+1)$-dimensional space of polynomials of degree $m + 2$. The 'unfurled' m-dimensional swallowtail is sent under this projection onto the usual m-dimensional swallowtail (formed by the polynomials with a multiple root). This map is one-to-one, except on the line of selfintersection of the swallowtail (for $n = 2$). Each point, except the origin, on this line has 2 pre-images on the unfurled swallowtail. Topologically, the unfurled swallowtail is homeomorphic to a euclidean space. This homeomorphism preserves all singularities of the usual swallowtail, except selfintersections. Thus, the 'lifting' of the usual swallowtail to the unfurled one (topologically equivalent to 'normalisation' in algebraic geometry) simplifies the topological structure and opens some loops based at points of selfintersection. The name 'open' or 'unfurled' swallowtail reflects these properties. As we will later see, open swallowtails control the singularities of ray systems at an obstacle. Here, however, we use m-dimensional Lagrangian varieties to define open umbrellas. Doing so, we forget the symplectic structure of the ambient $2m$-dimensional space. The conormal bundle of the m-dimensional unfurled swallowtail lives in the $4m$-dimensional symplectic space of the cotangent bundle of the space of polynomials. It is a Lagrangian variety of even dimension, $2m$, and is the image of a Lagrangian inclusion.

The singularities of the $2m$-dimensional open umbrella form a flag of submanifolds of even dimensions. The $2k$-dimensional subvariety in this flag is isomorphic to the $2k$-dimensional open umbrella.

Other (equivalent) definitions of the $2m$-dimensional open umbrella are:

1. The subvariety in the space of pairs of polynomials

$$\{e_{2m+1} + q_1 e_{2m-1} + \cdots + q_{2m} e_0,$$

$$p_{2m} e_{2m-1} - p_{2m-1} e_{2m-2} + \cdots - p_1 e_0\}$$

(where $e_s = x^s/s!$) formed by those pairs that have a common root of multiplicity at least $(m+1, m)$.

2. A component of the Lagrangian variety defined by the singular generating family

$$F(x, q, Q) =$$

$$= \int_0^x (Q_1 \xi^{m-1} + \cdots + Q_m)(\xi^{m+1} + q_1 \xi^{m-1} + \cdots + q_m)\, d\xi.$$

This family defines the Lagrangian variety

$$\left\{ p = \frac{\partial F}{\partial q}, \quad P = \frac{\partial F}{\partial Q}, \quad F_x = 0 \right\}$$

and we consider the component $x^{m+1} + q_1 x^{m-1} + \cdots + q_m = 0$.

Returning to ray systems, let us consider the usual situation: the system of normals to a hypersurface in a euclidean space, or the system of characteristics of a Hamilton— Jacobi equation corresponding to the given initial value of the unknown function, restricted to a hypersurface in the configuration space.

In these cases a generic boundary condition defines a Lagrangian submanifold transversally intersecting the hypersurface (defining the Hamilton—Jacobi equation). Hence the corresponding ray system is an immersed Lagrangian submanifold of the space of rays, and it has no open umbrella singularities.

As mentioned above, surfaces with nonzero Euler characteristic cannot be immersed in a symplectic space as Lagrangian submanifolds. However, Givental has constructed Lagrangian inclusions in the standard symplectic 4-space of surfaces with arbitrary Euler characteristic $\chi \le -2$. These can even be chosen to have no points of selfintersection, and so there is a homeomorphism onto the image of the inclusion [141]. Possibly these inclusions are not exact.

Consider an arbitrary Lagrangian inclusion of a surface L^2 in a symplectic manifold M^4. Givental has proved:

$$L \cdot L = \chi(L) + 2s + u,$$

where u is the number of open umbrella points, s is the number of points of selfintersection, χ is the Euler characteristic, and $L \cdot L$ is the selfintersection index of the

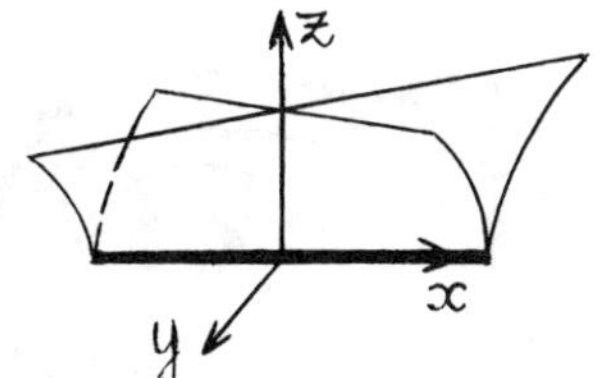

Figure 72: The folded umbrella

image of a fundamental cycle of L in $H_2(M)$. Selfintersections are counted with the signs defined by the orientation of L (if L is not orientable, the formula holds mod 2).

This formula suggests that a point of selfintersection or 2 open umbrella points form(s) a virtual antihandle. Indeed, in a neighborhood of any point of an embedded Lagrangian manifold we can attach a small Lagrangian handle, carrying either a point of selfintersection or 2 open umbrella points, the resulting Lagrangian variety also being embedded (in the case of umbrellas the handle destroys the orientation).

This construction may even preserve the exactness of the manifold: if the initial Lagrangian manifold is a projection of a Legendre manifold, the attachment can be performed in such a way that the resulting Lagrangian variety is also a projection of a Legendre variety.

The handles can be described in terms of the front corresponding to the exact Lagrangian variety in $\mathbf{R}^4 = T^*\mathbf{R}^2$. We first lift the manifold to a Legendre subvariety in $\mathbf{R}^5 = J^1(\mathbf{R}^2, \mathbf{R})$, and then project to the 3-space $J^0(\mathbf{R}^2, \mathbf{R}) = \mathbf{R}^3$. The image is the front variety.

The front of a smooth generic Lagrangian surface has as singularities semicubic cuspidal edges and swallowtails. The tangent plane is nowhere vertical (it is transversal to the fibers of the natural fibration $J^0(\mathbf{R}^2, \mathbf{R}) \twoheadrightarrow \mathbf{R}^2$). The selfintersections of the Lagrangian surface correspond to the vertical chords of the front for which the tangent planes to the front at the end points are parallel.

The open umbrella singularity of the Lagrangian variety corresponds to the 'folded umbrella' singularity on the front. The normal form of the *folded umbrella* is the surface in 3-space given by

$$y^2 = z^3 x^2.$$

This surface, which has a semicubic cuspidal edge and a line of selfintersection, is shown in Fig. 72. It is encountered in many problems in singularity theory.

Example 1. Consider a moving caustic (a generic 1-parameter family of caustics) in 3-space. The cuspidal edge of this caustic sweeps a surface, called a *bicaustic*. The generic singularities of bicaustics have been studied in [72], [142]. The folded umbrella is one such singularity.

Example 2. Consider the union of the tangent lines to a generic projective curve in space. This union is a *developable surface* whose cuspidal edge is the initial curve.

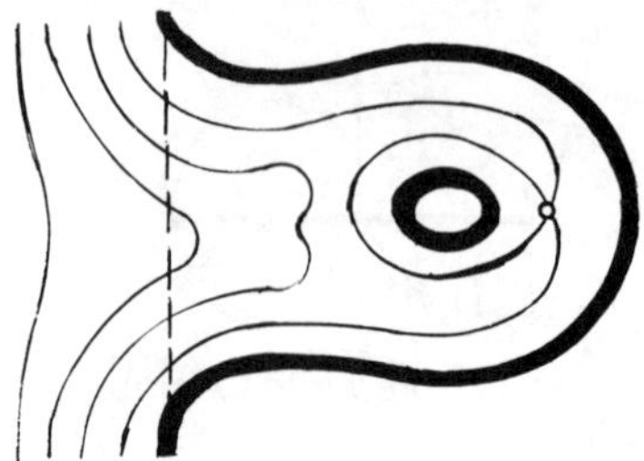

Figure 73: The front of a Lagrangian handle with one point of selfintersection

A generic curve may have isolated *flat points* (at which the torsion vanishes). The surface formed by the tangents has a folded umbrella singularity at such points ([143], [144]).

Example 3. Consider in 3-space a generic surface with an ordinary semicubic cuspidal edge. Consider a generic *folding map* of this 3-space (given by $(x, y, z) \mapsto (x^2, y, z)$ in suitable coordinates) The image of the surface has a folded umbrella singularity at the points of intersection of the cuspidal edge and the surface of the fold. This example and the topological equivalence of the folded umbrella and the Whitney umbrella are responsible for the name 'folded umbrella'.

Example 4. The folded umbrella provides some normal forms of generic singularities in Davydov's theory of slow motion in *relaxational systems* with one fast and two slow variables (see [145]–[147]).

A discussion of higher-dimensional generalisations of the folded umbrella may be found in [8].

Let us return to the attachment of Lagrangian handles. We start from a standard front with an ordinary cuspidal edge, defined locally by the equation $y^2 = x_1^3$ in the (x_1, x_2, y) coordinates of J^0. The new fronts with handles are described in Fig. 73 (one point of selfintersection of the Lagrangian immersion) and Fig. 74 (two open umbrellas of the Lagrangian inclusion). The fronts are symmetric with respect to the plane $y = 0$; the level lines of y on the fronts are shown in the figures. This symmetry implies that the only selfintersections of the Lagrangian variety that are possible occur at points of the front at which the tangent space is horizontal, that is, at the singular points of the level lines. Thus, Fig. 73 shows one point of selfintersection of the Lagrangian variety, and there is no such point in Fig. 74.

The above-mentioned construction of Givental suggests that elimination of pairs of open umbrellas of a Lagrangian inclusion of a surface is possible if the surface has a sufficient amount of handles.

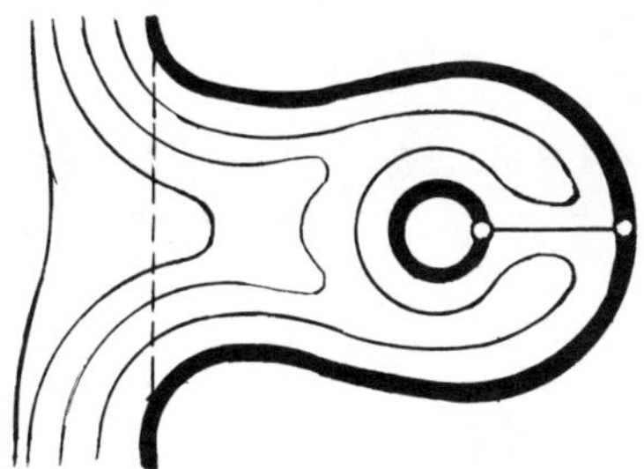

Figure 74: The front of a Lagrangian handle with two open umbrellas

Theorem. *A closed nonorientable surface whose Euler characteristic is a negative multiple of 4 admits a nonsingular Lagrangian embedding into the standard symplectic 4-space.*

(See [141] for a proof.)

It is not known whether the Klein bottle, the projective plane, and a surface of Euler characteristic -1 admit homeomorphic Lagrangian inclusions into the standard 4-space (inclusions with as singularities open umbrellas and selfintersections have been constructed by Givental in [141]).

Chapter 6

Projections of surfaces, and singularities of apparent contours

Projections are most common objects. For instance, we distinguish objects by the apparent contours of their projections to out retina surface, that is, by the critical values of the projection maps.

Singularities of projections are also crucial in the theory of bifurcation of equilibrium positions of dynamical systems. In this case the manifold that is projected lies in the product of the parameter space and the phase space of the dynamical system: it is formed by the equilibrium positions for all parameter values. The singularities of the projection of the manifold of equilibria into the parameter space are responsible for the bifurcations of the equilibria as the parameter values change.

The difference between the theory of singularities of projections and general singularity theory mainly depends on the smaller amount of available maps. The definition of equivalence of projections is also different from the general definition of (RL-) equivalence of maps. Namely, two projections (from submanifolds of the total spaces of two fibrations to the base spaces) are equivalent if there is a diffeomorphism from the first total space to the second that is fibered over a diffeomorphism from the first base space to the second and that sends the first submanifold onto the second.

In this chapter we will discuss the classification of singularities of projections up to this equivalence.

6.1 Singularities of projections from a surface to the plane

Consider a surface in 3-space. According to Whitney's classic theorem, the singularities of a generic projection are folds (along the critical lines) and Whitney tucks (at

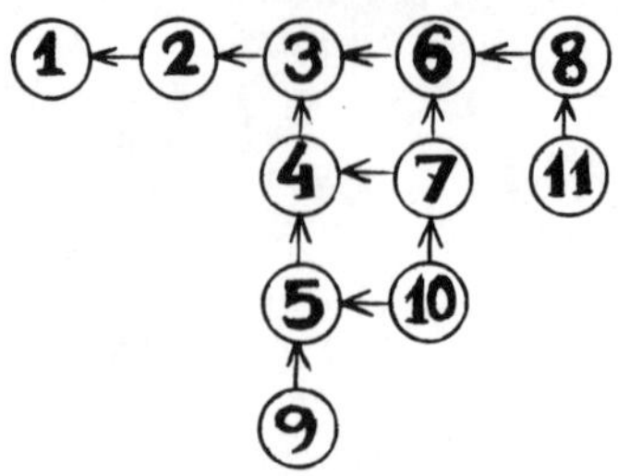

Figure 75: The hierarchy of germs of projections

isolated points). However, if we project a generic surface along a special direction, we may obtain more complicated singularities. In other words, a generic surface, seen from a particular direction, may have an unusual appearance.

The number of distinct patterns that we can obtain (by looking from outside the surface in an appropriate direction) is 14. Their study is based on the analysis of vector fields tangent to (generalised) swallowtails.

A *projection* of a surface M lying in the projective space $\mathbf{R}P^3$ from a point O not on M is a diagram $M \hookrightarrow \mathbf{R}P^3 \setminus O \twoheadrightarrow \mathbf{R}P^2$, where the left arrow is the inclusion map and the second is the projection sending a point to the line connecting it with O.

A germ of a projection is a similar diagram of germs. Two projections (germs) are *equivalent* if there is a commutative (3×2)-diagram whose rows are the projections (germs) and whose columns are the diffeomorphisms (germs).

The plane $\mathbf{R}P^2$ into which we project is called the *base* of the projection. Thus, equivalence means the existence of a diffeomorphism of the ambient 3-spaces that is fibered over a diffeomorphism of the base spaces.

The product maps $M \rightarrow \mathbf{R}P^2$ defined by equivalent projections are left—right equivalent (i.e., can be transformed to each other by diffeomorphisms of the source surface M and of the target space $\mathbf{R}P^2$).

The apparent contours (i.e., sets of critical points) of equivalent projections are diffeomorphic. Hence the classification of germs of projections up to equivalence of projections entails the classification of apparent contours up to diffeomorphism. However, a priori, diffeomorphism of apparent contours does not imply equivalence of the projections.

The hierarchy of germs of projections of generic surfaces is shown in Fig. 75 (it was computed by O.A. Platonova [9], and corrected by O.P. Shcherbak, [13], [148]). The figures denote the germs of the projections that are equivalent to the germs of the projections of the surfaces $z = f(x, y)$ along the x-axis. In the nonsingular case

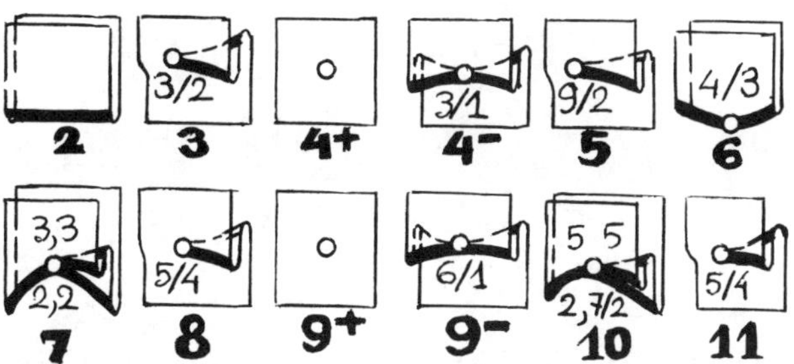

Figure 76: The singularities of the apparent contours of a generic surface

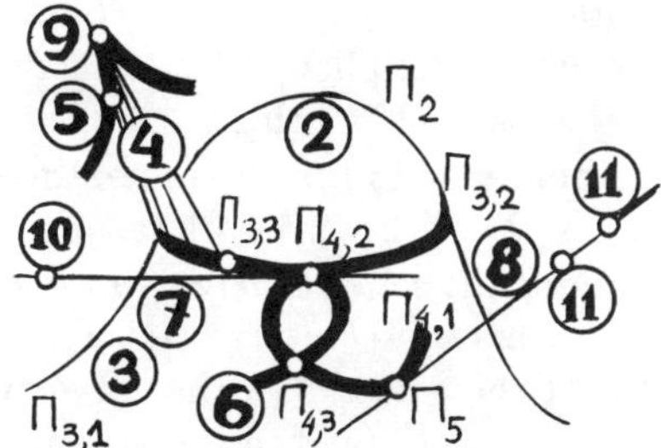

Figure 77: The tangential singularities of a surface in projective 3-space

1 the function f is x; the other functions f are:

type	$f(x, y)$	type	$f(x, y)$
2	x^2	7	$x^4 + x^2 y + xy^2$
3	$x^3 + xy$	8	$x^5 \pm x^3 y + xy$
4	$x^3 \pm xy^2$	9	$x^3 \pm xy^4$
5	$x^3 + xy^3$	10	$x^4 + x^2 y + xy^3$
6	$x^4 + xy$	11	$x^5 + xy$

Theorem. *A projection of a generic surface does not have germs that are inequivalent to the 14 germs above (for any choice of projection center (outside the surface)).*

The corresponding singularities of the apparent contours are shown in Fig. 76. The germ of a projection from a generic center is of type **1** at the generic points of the surface, of type **2** (a fold) at points on a certain line on the surface, and of type **3** (a Whitney tuck) at certain isolated points. Namely, tucks appear when the line of projection is tangent to the surface along an asymptotic direction.

In order to describe the other singularities, which are visible from specific directions only, we need the projective classification of points on a generic surface in projective 3-space (O.A. Platonova [9], E.E. Landis [25], Fig. 77).

A smooth *curve of parabolic points* $\Pi_{3,2}$ divides the surface into the *domain of elliptic points* Π_2 (no real tangent lines of order of tangency exceeding 2) and the *domain of hyperbolic points* $\Pi_{3,1}$ (two such lines, called *asymptotic lines*; their directions

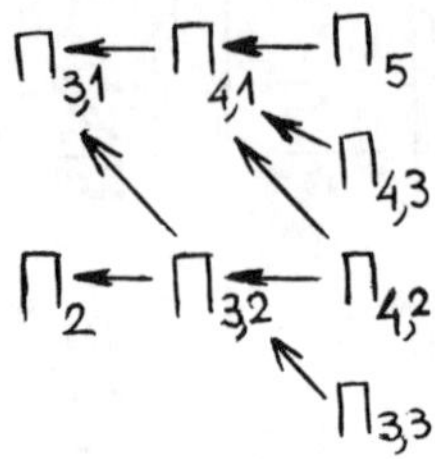

Figure 78: The hierarchy of tangent singularities

at the point of tangency are called *asymptotic directions*). In this notation, the first subscript indicates the maximal multiplicity of intersection of a tangent line and the surface (the order of tangency is equal to this multiplicity minus 1).

In the domain of hyperbolic points lies the (smoothly immersed) *curve of inflections of the asymptotic lines* $\Pi_{4,1}$ (the order of tangency of an asymptotic tangent exceeds 2). This curve contains the set of isolated *points of bi-inflection* Π_5 (tangency of order 4) and the set of *points of selfintersection* $\Pi_{4,3}$ (both asymptotic tangents have tangency of order 3). At the *points* $\Pi_{4,2}$ the curve of inflections is (simply) tangent to the curve of parabolic points (at these points the only asymptotic direction is tangent to the curve of parabolic points). Finally, the curve of parabolic points contains isolated *points of type* $\Pi_{3,3}$, described below.

The resulting hierarchy of classes of points on the surface is shown in Fig. 78 (classes in the same column have equal codimension).

A surface projectively dual to our generic smooth surface has a semicubic cuspidal edge, at the points corresponding to $\Pi_{3,2}$ and $\Pi_{3,3}$, and a swallowtail, at the points of $\Pi_{4,2}$.

The p-jets of the surface at these points are reducible by a projective transformation to the following normal forms (see [9]):

class	normal form	restrictions	p	codim
Π_2	$x^2 + y^2$	—	2	0
$\Pi_{3,1}$	$xy \pm x^3 + y^3$	—	3	0
$\Pi_{3,2}$	$y^2 + x^3 + xy^3 + ax^4$	—	4	1
$\Pi_{4,1}$	$xy + y^3 + x^4 + hx^3y$	—	4	1
$\Pi_{4,2}$	$y^2 + x^2y + ux$	$u \neq 0, \frac{1}{4}$	4	2
$\Pi_{4,3}$	$xy + x^4 + ax^3y + bxy^3 \pm y^4$	—	4	2
Π_5	$xy + y^3 \pm x^3y + \phi_5$	$\phi_5(0,y) \neq 0$	5	2
$\Pi_{3,3}$	$y^2 + x^3 + ax^4 + by^5 \pm xy^4 + x^2\phi_3$	—	5	2

Here ϕ_r denotes a homogeneous polynomial of degree r in x and y.

Theorem. *Any p-jet of a generic surface is projectively equivalent to that of a surface* $z = f(x, y)$ *with* f *one of the above normal forms.*

Remark. For a generic algebraic surface of degree d in $\mathbf{C}P^3$ the numbers of points in classes of codimension 2 and the degrees of the curves formed by the points of classes of codimension 1 depend on d only. These numbers are (according to [149], [26], [150]):

$$\begin{array}{cccc} \Pi_5 & \Pi_{4,3} & \Pi_{4,2} & \Pi_{3,3} \\ 5d(d-4)(7d-12) & 5d(7d^2-28d+30) & 2d(d-2)(11d-24) & 10d(d-2)(7d-16) \end{array}$$

$$\begin{array}{c|cc} \text{curve} & \Pi_{3,2} & \Pi_{4,1} \\ \hline \text{degree} & 4d(d-2) & d(11d-24) \end{array}$$

Corollary. *A real algebraic surface of sufficiently high, odd degree in* $\mathbf{R}P^3$ *has at least one real point of bi-inflection of asymptotic lines, at least one real point of inflection of both asymptotic lines, and at least one curve of parabolic points.*

Remark. For similar results concerning hypersurfaces in $\mathbf{C}P^4$ see [151] and [152].

We return to the description of projections from nongeneric centers (Fig. 77). The singularities **4** and **6** occur when projecting from points on certain surfaces (formed by the asymptotic tangent lines at the parabolic points and at the points of inflection of the asymptotic lines, respectively). Thus, in order to see the singularity **4**, one has to look at a generic surface from a generic point on an asymptotic tangent line at a generic parabolic point. In order to see the singularity **6**, one has to choose the projection center on an asymptotic tangent line of order 3.

The singularities **5**, **7**, and **8** occur when projecting from points on certain curves. Namely, in order to see the singularity **5**, one has to look at a generic surface from a generic point on the cuspidal edge of the surface formed by the asymptotic tangent lines at the parabolic points (this surface is developable and has a cuspidal edge).

The singularity **7** occurs when projecting from a generic point on an asymptotic tangent line tangent to the curve of parabolic points (that is, issuing from $\Pi_{4,2}$). The singularity **8** can be seen from generic points of asymptotic tangent lines of order 4.

The, rarest, singularities **9**, **10**, and **11** can be seen from isolated points. Namely, the singularity **9** can be seen from the cusp point of the cuspidal edge of the developable surface described above (the corresponding asymptotic tangent line issues from $\Pi_{3,3}$). The singularity **11** can be seen from two 'focal' points on an asymptotic tangent line of order 4. The singularity **10** occurs when projecting from a special point on the asymptotic line tangent (at $\Pi_{4,2}$) to the curve of parabolic points.

This classification of projections is as follows related to vector fields tangent to swallowtails.

Consider the intersection of one line of projection and the surface as a 0-dimensional subvariety of the line. For a generic line the intersections are transversal, but at singularities intersection points collide. Thus, each singularity of the projection of a surface in 3-space to a plane defines a 2-parameter deformation of a 0-dimensional subvariety of a line (consisting of μ collided points, where μ is the multiplicity of the intersection of the line of projection and the surface at a point of tangency of both). The 2 parameters define a line in the projection bundle: the base of the deformation is the (germ of the) projective plane $\mathbf{R}P^2$ of lines issuing from the projection center. This 2-parameter deformation, as any other, is equivalent to a deformation induced from a versal deformation. Hence it defines a map from the base of our deformation into the base of this versal deformation (that is, into a space in which a generalised swallowtail, the discriminant of the singularity of type A_μ, lives). For instance, for $\mu = 3$ we obtain a surface containing the vertex of the ordinary swallowtail in $\mathbf{R}^3$.

A diffeomorphism of the base space of the versal deformation preserving the (generalised) swallowtail transforms our surface into a new surface. This new surface defines another 2-parameter deformation of the μ-multiple point, hence a new projection. This new projection is equivalent to the initial projection in our sense of equivalence of projections (for more details see [98]).

Thus, in order to reduce a projection to normal form it suffices to reduce to normal form the corresponding surface in the space in which the (generalised) swallowtail lives, using a diffeomorphism preserving the swallowtail. Knowledge of the vector fields tangent to the swallowtail allows one to reduce to normal form various objects in the space of the swallowtail, using a diffeomorphism preserving the swallowtail (one could use standard homotopy and standard quasihomogeneous methods, or spectral sequence techniques). For more details see [98], [1].

The projections of a surface from points of the 3-space form a 3-parameter family of maps. In generic 3-parameter families of maps from a surface to a plane, certain singularities distinct from the list **1–11** occur. According to computations by V.V. Gorjunov, these are equivalent to the germs at the origin of the projections along the x-axis of the surfaces

$$z = x^4 \pm x^2 y^2 + xy^2, \quad x^5 + x^3 y + x^2 y, \quad x^6 + ax^4 y + x^3 y + xy.$$

The theorem of Platonova and Shcherbak shows that these singularities do not occur when projecting from an arbitrary point a generic surface in 3-space. The reason for this is the strong interdependence of projections from points on the same line of projection: the 3-parameter families occurring in the projection problem are rather special.

Consider, for instance, the last singularity in the list, having $\mu = 6$. A generic projection has isolated Whitney tucks, where $\mu = 3$. Hence a generic 3-parameter family has isolated points at which $\mu = 6$. But the multiplicity of the intersection of the line of projection and the surface is independent of the choice of projection center on this line. Hence isolated centers at which the multiplicity jumps are impossible. Indeed, the space of tangent lines is 3-dimensional. Each increment of the order of tangency by 1 imposes one more restriction on the tangent line. Hence the maximal order of tangency is 4 (the maximal multiplicity of intersection is 5). Consequently, there does not exist a projection center in 3-space for which a singularity with $\mu = 6$ occurs (provided the projected surface is generic).

The same reasoning shows that a singularity with $\mu = 6$ cannot occur when projecting generic surfaces along extremals of an arbitrary variational problem (and even along the curves of an arbitrary 4-parameter family of curves in a 3-dimensional manifold).

Similar reasons explain the nonoccurrence of the other two codimension-3 singularities in Gorjunov's list (this reasoning is due to O.P. Shcherbak): they rather reflect the properties of the line of projection than those of the center. The singularity $x^5 + x^3y + x^2y$ would only be possible for a projection from a point on an asymptotic tangent line having at a parabolic point order of tangency 4 with the surface. However, a generic surface does not have such tangents.

The singularity $x^4 \pm x^2y^2 + xy^2$ would only be possible for a projection from a point on an asymptotic tangent line having, at a point at which the curve of parabolic points is nonsmooth, order of tangency 3 with the surface. However, a generic surface does not have such points.

The singularities of projections by bundles of parallel lines were classified in [98] (10 types; the 4 singularities of types **9, 10, 11** do not occur). This list coincides with the list of singularities of generic 2-parameter families of projections (Gorjunov), and with the list of singularities whose left—right codimension does not exceed 2 (see [153], [133]).

When looking at a generic surface for one of the most singular points (**9, 10, 11**), the shape that we see will change when we displace our point of view a little. The resulting patterns are given in Figs. 79–82. In the middle of each figure the bifurcation diagram formed by the centers of the nongeneric projections is shown. The projections corresponding to centers that belong to different strata of the bifurcation diagram and its complement are shown along the sides of the figures.

6.2 Singularities of projections of complete intersections

The classifications of singularities of various objects show that the algebraically most natural classifications are those of simple objects, that is, of objects without moduli. Thus, the classification of the simple critical points of functions, and that of the simple singularities of hypersurfaces (as well as that of the simple singularities of Lagrangian and Legendre maps, and of the simple singularities of caustics and wave fronts), leads to the list A_μ, D_μ, E_6, E_7, E_8 of Dynkin diagrams without multiple edges (angles of 120° between the nonorthogonal simple roots), see [2]. The classification of the simple critical points of functions on a manifold with boundary leads to the same list, completed by the diagrams B_μ, C_μ, F_4 (angles of 135° are admitted).

Hence in any problem it is natural to look for the simple objects. In doing this it is useful to directly study the bifurcation diagrams, since they play the role of fingerprints of singularities. Thus, the recognition of the simple Lie algebras in the list of simple boundary singularities was due to the topological equivalence of the bifurcation diagrams occurring in both theories (see [3]).

The classification of the simple singularities of projections $V \hookrightarrow E \twoheadrightarrow B$ of subvarieties V of the total space E of fibrations $E \twoheadrightarrow B$ into the base spaces B, up

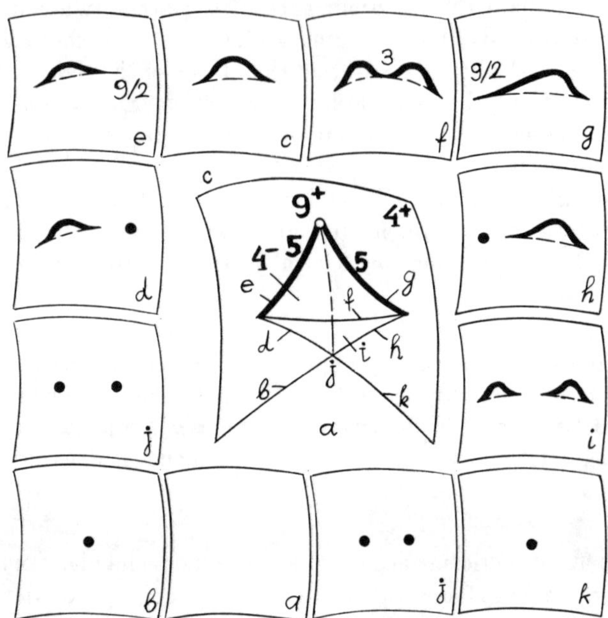

Figure 79: The bifurcations of the projection $x^3 + xy^4$

to equivalence (i.e., up to a diffeomorphism of E that is fibered over B) is due to V.V. Gorjunov [131].

In his theory, V is not assumed to be a smooth manifold, but rather a complete intersection (defined by equations whose total number equals the codimension of V in E). These restrictions are natural in the theories of metamorphoses and cobordism of projections. Indeed, in these theories the singular projections occur for bifurcation values of the parameters on which the projection depends. This situation is described by a diagram

$$V' \hookrightarrow E' \twoheadrightarrow B' \twoheadrightarrow P,$$

where P is the parameter space and V' is a smooth submanifold in the total space of the smooth fibration $E' \twoheadrightarrow B'$.

A fiber of the fibration over P defines a projection $V_p \hookrightarrow E_p \twoheadrightarrow B_p$. The variety V_p becomes singular for bifurcation values of the parameter p.

We define a *suspension of a projection* $V \hookrightarrow E \twoheadrightarrow B$ to be an inclusion of E into the space of a larger fibration with the same base space B (as that of the subfibration). *Stable equivalence of projections* means equivalence of suitable suspensions of them. A *projection 'on'* is a projection for which the dimension of the base space B does not exceed that of the projected variety V.

The projection from V to B defines a family of subvarieties in the fibers of the fibration $E \twoheadrightarrow B$ (the family of intersections of V with the fibers). A germ of a projection $V \hookrightarrow E \twoheadrightarrow B$ may be regarded as a deformation of the variety V_0 in a

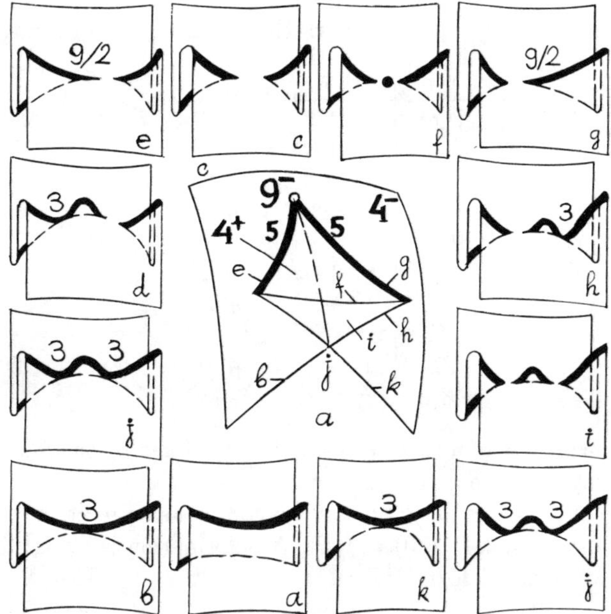

Figure 80: The bifurcations of the projection $x^3 - xy^4$

distinguished fiber. The codimension of V_0 in this fiber is equal to the codimension of V in E.

Theorem (see [131]). *A simple germ of a projection 'on' is fiberwise stably equivalent to a deformation of either a hypersurface, a curve in 3-space, or a (multiple) point in the plane.*

In each of the three cases, Gorjunov has listed all simple singularities. Parts of these lists coincide with the lists of simple singularities in other classification problems. For instance, he has proved the following:

Theorem. *A deformation of a curve in $\mathbf{C}^3$ defines a simple germ of a projection if and only if it is a versal deformation of a simple singularity of a curve in $\mathbf{C}^3$.*

The list of simple singularities of nonplanar curves in $\mathbf{C}^3$, defined by 2 equations, has been published by M. Giusti [154]. The hierarchy of these singularities is shown in Fig. 83.

The list[1] contains an infinite sequence

$$S_\mu: \quad x^2 + y^2 + z^{\mu-3} = yz, \qquad \mu \geq 5,$$

---

[1] Simple singularities of complete intersections may be viewed as simple singularities of hypersurfaces in supermanifolds (M. Kasarian, 1990).

and 10 exceptional curves $f_1 = f_2 = 0$, with the f's as in the following table:

type	f
T_7	$x^2 + y^3 + z^3,\ yz$
T_8	$x^2 + y^3 + z^4,\ yz$
T_9	$x^2 + y^3 + z^5,\ yz$
U_7	$x^2 + yz,\ xy + z^3$
U_8	$x^2 + yz + z^3,\ xy$
U_9	$x^2 + yz,\ xy + z^4$
W_8	$x^2 + z^3,\ y^2 + xz$
W_9	$x^2 + yz^2,\ y^2 + xz$
Z_9	$x^2 + z^3,\ y^2 + z^3$
Z_{10}	$x^2 + yz^2,\ y^2 + z^3$

The *versal deformation* of a variety at a point is defined by the following construction (for its motivation see, e.g., [28]).

A deformation of the germ $\{x : f(x) = 0\}$ of codimension m at the origin is defined by a germ $F(x, \lambda)$ such that $F(x, 0) = f(x)$. A deformation is *versal* if the images of the initial velocities of the deformation,

$$g_i(x) = \left. \frac{\partial F}{\partial \lambda_i} \right|_{\lambda=0},$$

in the quotient module

$$\mathcal{O}^m / \left\langle \frac{\partial f}{\partial x_i}, f_i e_j \right\rangle$$

generate this module over $\mathbf{C}$. Here $\mathcal{O} = \mathbf{C}[[x]]$ is the $\mathbf{C}$-algebra of Taylor series at the origin, $\mathcal{O}^m = \oplus \mathcal{O} e_i$ is the free $\mathcal{O}$-module with free basis $\{e_i\}$, and the angular brackets indicate the submodule generated by the elements between it.

Example. Consider the curve S_5,

$$f_1 = x^2 + y^2 + z^2, \qquad f_2 = yz.$$

As a versal deformation we may take

$$F_1 = f_1 + \lambda_1, \qquad F_2 = f_2 + \lambda_2 + \lambda_3 x + \lambda_4 y + \lambda_5 z.$$

The projection of the surface $F_1 = F_2 = 0$ to the λ-space has a simple singularity at the origin.

In this way Giusti's list (given above) is included in Gorjunov's list of deformations of nonplanar curves.

Comparison of Giusti's list and Gorjunov's list shows that the subvarieties V occurring in the list of simple singularities of projections $V \hookrightarrow E \twoheadrightarrow B$ that are not stably equivalent to projections of hypersurfaces, are in fact *smooth* submanifolds of E. For the other simple projections the complete intersection V need not be smooth.

Gorjunov's list of simple projections of complete intersections is rather long, especially in the real domain. (Also, it does not coincide with any other known list of simple objects.)

Here we reproduce that part of his list that corresponds to projections 'on' the complex line.

Theorem (see [131], [130]). *1) For a simple germ of a projection on a line the projected manifold is (up to stable fibered equivalence) either a hypersurface, a (singular) curve in a 3-space, or a (multiple) point on a plane or in a 3-space.*

2) A simple germ of a projection of a hypersurface in the space of the fibration $(x, u) \mapsto u$ is reducible by a fibered holomorphic diffeomorphism $(x, u) \mapsto (h(x, u), v(u))$ to the projection on the u-axis of one of the hypersurfaces in the following list

type	hypersurface
$A_\mu, \mu \geq 0$	$u + x_1^{\mu+1} + q = 0$
$D_\mu, \mu \geq 4$	$u + x_1^2 x_2 + x_2^{\mu-1} + q = 0$
E_6	$u + x_1^3 + x_2^4 + q = 0$
E_7	$u + x_1^3 + x_1 x_2^3 + q = 0$
E_8	$u + x_1^3 + x_2^5 + q = 0$
$B_\mu, \mu \geq 2$	$x_1^2 + u^\mu + q = 0$
$C_\mu, \mu \geq 3$	$x_1^\mu + u x_1 + q = 0$
F_4	$x_1^3 + u^2 + q = 0$

where q denotes the sum of the other variables ($q = x_2^2 + \cdots + x_n^2$ for A_μ, B_μ, C_μ, F_4, while for D_μ and E_μ, $q = x_3^2 + \cdots + x_n^2$). The adjacency diagram is shown in Fig. 84.

3) The simple projections of the germs of curves in 3-space onto a line form 2 infinite sequences (one of them having 2 indices); they are reducible to projections on the u-axis of the germs at the origin of the following curves:

type	restrictions	curve	
$C_{k,l}$	$2 \leq k \leq l$	$xy = 0,$	$x^k + y^l + u = 0$
F_{2k+1}	$2 \leq k$	$x^2 + y^3 = 0,$	$y^k + u = 0$
F_{2k+4}	$1 \leq k$	$x^2 + y^3 = 0,$	$xy^k + u = 0$

The adjacency diagram is shown in Fig. 85.

4) The simple projections from complete intersections to the line form 2 infinite sequences (one of them having 2 indices) and 7 exceptional singularities; they are equivalent to projections to the u-axis of the following germs at the origin of multiple *points in the (x, u)-plane or in the (x, y, u)-space:*

type	restrictions	multiple point
B_μ	$\mu \geq 0$	$x = u^\mu = 0$
$X_{k,l}$	$2 \leq k \leq l$	$x^k + u = x^l = 0$
U_μ	$4 \leq \mu$	$x^2 + u^{\mu-2} = ux = 0$
V_6		$x^2 + u^2 = 0$
V_7		$x^3 + u^2 = ux = 0$
$\Gamma^{a,b}_{a+b+3}$	$a = 2, \quad b = 2 \; or \; 3$ $a = 3, \quad b = 2, 3, \; or \; 4$	$xy = x^2 + y^a = u + y^b$

The adjacencies are shown in Fig. 86.

This classification leads to the following conclusions.

1. The boundary singularities. The simple singularities of projections of hypersurfaces are classified (up to complex stable equivalence) by the Weyl groups $A_\mu, \ldots, F_4$, that is, by the same list that classifies the boundary singularities [3]. The usual (i.e. not boundary) singularities correspond to projections of *smooth* hypersurfaces.

The equivalence relation in the theory of boundary singularities of hypersurfaces is 'wider' than that in the theory of singularities of projections of hypersurfaces: in the first case the equivalences are diffeomorphisms preserving the boundary (one hypersurface), while in the second case the equivalences preserve a fibration (into hypersurfaces parallel to a given hypersurface).

The coincidence of both lists is an a priori unexpected result. They also coincide with the list of simple boundary singularities of functions (and with those of the simple boundary Lagrangian and Legendre singularities). Infinitesimal 'explanations' of these coincidences (using quasihomogeneity arguments) are presented in § 2 of [133].

Remark. The unimodular and the bimodular boundary singularities of functions are classified in [3], [155], [156]. The resulting lists have not yet been identified with other interesting classifications.

The theory of Lagrangian boundary singularities leads to an interesting 'Lagrangian duality', interchanging a function on the ambient space and its restriction to the boundary (this version of Lagrange's 'multiplier rule' is due to I.G. Shcherbak [157]).

Let $f(x, y)$ $(x \in \mathbf{R}, y \in \mathbf{R}^{n-1})$ be a function germ at a critical point 0 of the n-space with 'boundary' $x = 0$. Its *Lagrangian dual function* is defined by the formula $f^*(z, x, y) = zx + f(x, y)$, where the equation of the 'dual boundary' is $z = 0$. The restriction of f^* to the boundary is f, while the restriction of f to the boundary is stably equivalent to f^*. Also, f^{**} is stable equivalent to f, regarded as a function on a manifold with boundary.

The bifurcation diagram of the zeros of f and that of f^* are diffeomorphic, but the components (corresponding to degeneration of the 0-level variety and to nontransversality of it on the boundary) are interchanged.

Thus, boundary singularities may be viewed as 'products' of ordinary singularities:

$$C_k \sim (A_1, A^{k-1}), \quad B_k \sim (A_{k-1}, A^1), \quad F_4 \sim (A_2, A^2)$$

(the index, equal to the multiplicity, is written in the sup if it corresponds to a singularity of the restriction to the boundary, and in the sub if it corresponds to a singularity in the ambient space).

The 'product' is not uniquely determined by the factors, and 'multiplication' is not always possible. For instance, it is impossible to multiply A_2 and A^3 (and, more generally, A_{2k} and A^p when $p > 2k$).

The products of A_p and A^p form an m-parameter family of singularities, where $m = [(p-1)/2]$:

$$F_p^p = x^2 + y^{p+1} + xa(y),$$

where $a(y) = a_1 y^{p-1} + \cdots + a_m y^{p-m}$.

The products of simple singularities have been classified by I.G. Shcherbak [157]. The Dynkin diagram of a *simple* product decomposes into the diagrams of the factors when a double edge (pointing towards the diagram of the restriction of the function to the boundary) is deleted. The modality of a product is majorated by the sum of the modalities of the factors, at least in the examples.

2. Projections of curves. The notation of the sequences C and F reflects the intrinsic relation of these singularities to the singularities C_k and F_4 of hypersurfaces. Indeed, for these singularities V.V. Gorjunov has proved theorems about the geometry of bifurcation diagrams that are similar to those in the ordinary theory. He has also defined the Dynkin diagrams of these singularities (see Fig. 88). Thus, the exceptional Lie algebra F_4 is the ancestor of a sequence F_k of related objects.

3. The intersection form. For defining the intersection form of projections onto the line, $V_0 \hookrightarrow E \twoheadrightarrow \mathbf{C}$, we use the same method as in the case of boundary singularities.

Fix in the space E a small ball B_r of radius r and center 0, and choose a small (with respect to r) positive ϵ. Let the complete intersection V_0 be defined by the equation $f = 0$ in E (Fig. 87). A *Milnor fiber* of a projection is a nonsingular level manifold $V_\xi = B_r \cap f^{-1}(\xi)$, $|\xi| = \epsilon$, with ξ generic. This manifold intersects the fiber of the fibration $E \twoheadrightarrow \mathbf{C}$ over 0 transversally along a manifold of complex codimension 1 in V_ξ, denoted by V_ξ^{o}.

Consider the 2-fold covering $\tilde{V}_\xi \to V_\xi$ of V_ξ, ramified along V_ξ^{o} (the 'complexification' of the 'boundary' is the "2-fold ramified covering"). Consider the integer homology group $\tilde{H}$ of $\tilde{V}_\xi$ of middle dimension. Permutation of the branches of the covering acts on $\tilde{H}$ as an involution. Let H^- be the anti-invariant part of $\tilde{H}$. The intersection form on $\tilde{H}$ induces on H^- a bilinear form, called the *bilinear form of the projection*.

4. The Dynkin diagram. If the complex dimension of V_ξ is even, the intersection form is symmetric. In this case the Dynkin diagram is defined as for boundary singularities. Namely, the anti-invariant space is generated by the *long roots* (a long root is the difference of two pre-images of a vanishing cycle on V_ξ) and the *short roots* (a short root is formed by the pre-images of a relative cycle $V_\xi \bmod V_\xi^o$, that is, of a vanishing semi-cycle; for details see [3]).

If the complex dimension of V_ξ is odd (in our case, when V_ξ is a curve, in the situations F_k or $C_{k,l}$), the definitions of short and long roots are not so obvious. In the case of boundary singularities this difficulty is irrelevant, since we then have a stabilisation procedure, transforming a function with skewsymmetric intersection form to a function with symmetric intersection form. For projections such a procedure does not exist (or is unknown), hence we define short and long roots by the following construction.

Consider both the anti-invariant H^- and the invariant part H^+ of the group $\tilde{H}$. The group H^- can be naturally projected onto the quotient group $\tilde{H}/H^+$. An anti-invariant element will be called a *short element* if its image in $\tilde{H}/H^+$ is not divisible by 2, and a *long element* otherwise.

The Dynkin diagram of a skewsymmetric form is a graph whose vertices represent the basic vectors. Two vertices are joined by a number of edges (this number is equal to the value of the form on the pair of basic vectors). The edges are oriented (making the above mentioned values positive). An edge joining a long vertex with a short one is labelled with the sign $<$, pointing towards the short vertex.

When the graph is a tree (possibly with multiple edges), the orientations of the edges are not indicated in the diagram (they may be arbitrary, possibly reversing basic vectors).

With these stipulations, the boundary singularities admit the usual Dynkin diagrams $A_k, \ldots, F_4$.

Theorem (see [131]). *In suitable bases the intersection forms of the projections F_k and $C_{k,l}$ are defined by the Dynkin diagrams of Fig. 88.*

5. Versal deformations of projections. The bases mentioned above may be chosen to be distinguished, as in the usual theory of critical points of functions. However, we have to adapt to our situation the definition of distinguished basis from the theory of boundary singularities [3].

First we need some notation. Consider the projection $(x, u) \mapsto u$ from the complete intersection V (defined in E by the equation $f(x, u) = 0$) to the base $B = \{u\}$. A *deformation* of this projection is defined by the equations $F(x, u, \lambda) = 0$, where $F(x, u, 0) \equiv f(x, u)$ and λ belongs to a neighborhood of the origin in a finite-dimensional space.

A *versal deformation* may be chosen to be

$$F = f + \sum \lambda_i g_i(x, u),$$

where the *deformation velocities* g_i generate over $\mathbf{C}$ the quotient space

$$Q_f = \{\alpha\} \bigg/ \bigg\{ \sum f_i \beta_i + \sum \gamma_j \frac{\partial f}{\partial x_j} + \sum \delta_s \frac{\partial f}{\partial u_s} \bigg\},$$

where α, β, γ, δ are holomorphic in a neighborhood of the space $\{(x,u)\}$ (α and each component β_i being vector functions with number of components equal to the number of components of f, δ_s and γ_j being scalar functions, and $\delta_s = \delta_s(u)$ being independent of x).

Remark. The nominator is the tangent space of the space of deformations of f. The first summand in the denominator is generated by the (infinitesimal) deformations of f that leave the variety $f = 0$ unchanged. The other summands are generated by the infinitesimal deformations of E that are fibered over B.

Example. The curve $C_{2,2}$ in $\mathbf{C}^3$ is defined by $f_1 = f_2 = 0$, where $f_1 = xy$, $f_2 = x^2 + y^2 + u$. Hence $\dim_{\mathbf{C}} Q_f = 3$, and a 3-parameter versal deformation is defined by

$$F_1 = f_1 + \lambda_1 + \lambda_2 x + \lambda_3 y, \qquad F_2 = f_2.$$

6. Vanishing cycles and semicycles. Having chosen a versal deformation, defined by the map F, we define the maps

$$F_\lambda : (x,u) \mapsto F(x,u,\lambda), \qquad F_\lambda^{\mathrm{o}} : (x,0) \mapsto F(x,0,\lambda).$$

The set of critical values of these maps are hypersurfaces in the space of values of F. Suppose that a point ξ in this space is not a critical value of the map F_λ. The variety $V_{\lambda,\xi}$ defined by the equation $F_\lambda = \xi$ in the space $\{(x,u)\}$ is then nonsingular. If ξ is not a critical value of F_λ^{o}, then $V_{\lambda,\xi}$ transversally intersects the hyperplane $u = 0$ along a smooth hypersurface $V_{\lambda,\xi}^{\mathrm{o}} \subset V_{\lambda,\xi}$.

Starting with the pair $(V_{\lambda,\xi}, V_{\lambda,\xi}^{\mathrm{o}})$, we define the intersection form in the same manner as in point *3* above for the pair $(V_\xi, V_\xi^{\mathrm{o}})$, which corresponds to $\lambda = 0$. The points (λ, ξ) and $(0, \xi)$ may be connected by a path that avoids the hypersurface formed by the critical pairs. Over this path the pairs form a smooth fibration. We can choose a trivialisation of this fibration over this path that fixes a neighborhood of the boundaries (of balls in the space $\{(x,u)\}$). This trivialisation defines isomorphisms of the homology of the 2-fold branched covering and the group of anti-invariant homology classes H^- for different values of (λ, ξ), and of the intersection forms on the middle homology groups.

We will use these isomorphisms to define vanishing cycles and semicycles in complete intersections on a manifold with boundary.

Choose a generic point λ (near the origin) and consider a generic complex straight line passing through a noncritical point ξ. The pre-image of this line under the map F_λ is a smooth manifold, whose dimension is that of the manifold $V_{\lambda,\xi}$ plus 1. This manifold contain the hypersurface $u = 0$ (which is the pre-image of the same line under the map F_λ^{o}). Assume that our line transversally intersects the set of critical values of the map F_λ in a points, and the set of critical values of F_λ^0 in a^0 points (all intersections being generic). Move the point ξ to these critical points along nonintersecting paths (this is always done in the definition of distinguished bases, see [44] and Fig. 89).

When ξ approaches the a points of the first kind, Picard—Lefschetz cycles vanishe on $V_{\lambda,\xi}$. When it approaches the a^0 points of the second kind, semicycles, realising relative homology classes belonging to $H(V_{\lambda,\xi}, V_{\lambda,\xi}^0)$, vanish (see [3]). These $a + a^0$ cycles and semicycles generate the whole relative homology group, but they may be dependent.

Any vanishing cycle (semicycle) defines an anti-invariant cycle in $H_{\lambda,\xi}^-$. These cycles are called the *distinguished cycles* (a vanishing cycle defines a long distinguished cycle, a vanishing semicycle defines a short one). The Dynkin diagrams of Fig. 88 are the diagrams of intersection forms in the bases formed by some distinguished cycles (superfluous distinguished cycles are neglected).

Remark 1. These superfluous distinguished cycles are a manifestation of the following general phenomenon: the natural generalisation of root systems (corresponding to complete intersections of hypersurfaces) should live not in the homology of the complete intersection itself, but rather in certain spaces associated with a flag of submanifolds of all dimensions, each of them being a hypersurface in the previous one.

Consider a complete intersection (or its nonsingular Milnor fiber) as a level hypersurface of a function on a complete intersection of dimension higher by one. The critical points of this function define the vanishing cycles in the middle homology of the Milnor fiber, and even define the systems of distinguished cycles. However, generically these systems do not form bases, since the number of critical points is larger than the middle Betti number of the fiber.

Hence we consider the free abelian group generated by the vanishing cycles of the system of distinguished cycles, together with a homomorphism into the homology of the fiber, sending a vanishing cycle to its homology class.

In order to study the kernel of this homomorphism we have to investigate the homology of the complete intersection of dimension higher by one, on which our fiber is a level hypersurface.

Repeating the previous construction, we obtain a 'resolvent', or tower, each floor of which is equipped with an intersection form, a monodromy, a mixed Hodge structure, etc. Unfortunately, neither the geometry nor the algebra of this situation has yet received the treatment they deserve.

Problem. Find the generalisation of the crystallographic Coxeter groups to this tower situation (the answer would include the list of towers corresponding to simple projections of curves).

Remark 2. For projections of complete intersections to the line there is another type of 'distinguished basis' (leading, in general, to different Dynkin diagrams).

Consider the projection to the line as a function u. For generic λ, the restriction of u to the manifold $V_{\lambda,0}$ (in the notation above) will be a Morse function with μ nondegenerate critical points and μ distinct critical values. The number of critical values is equal to the number of spheres in the bouquet to which the space $V_{\lambda,0}/V_{\lambda,0}^{\circ}$ is homotopy equivalent. Hence this number equals the rank of the group H^- of anti-invariant homology classes of the 2-fold covering $\tilde{V}_{\lambda,0} \to V_{\lambda,0}$, ramified along $V_{\lambda,0}^{\circ}$. If the projection is simple, $\mu = \nu + 1$, where ν is the dimension of the base space of the minimal versal deformation.

When u moves from a noncritical value ($u = 0$) to the μ critical values along μ distinguished paths in the complex line $\{u\}$, the level hypersurface on $V_{\lambda,0}$ degenerates. At the moment of degeneration, μ semicycles, and hence μ homology classes of $V_{\lambda,0}/V_{\lambda,0}^{\circ}$, vanish. The corresponding anti-invariant cycles in the homology of the 2-fold covering $\tilde{V}_{\lambda,0} \to V_{\lambda,0}$, ramified along $V_{\lambda,0}^{\circ}$, form a distinguished (in the new sense) basis of μ short cycles in H^-.

For the singularities $A_\mu, \ldots, E_8$ this construction leads to the usual Dynkin diagrams. But for B_μ the result is unusual: in the symmetric case the diagram splits into points (our basis is formed by the roots e_i, so that the other roots are $\pm e_i$, $\pm e_i \pm e_j$).

Other consequences of the classification of simple projections are a description of the geometry of the corresponding bifurcation diagrams. We first recall some general properties of the latter.

6.3 Geometry of bifurcation diagrams

One of the simplest bifurcation diagrams is the semicubic parabola, formed in the (a, b)-plane by the points for which the polynomials $x^3 + ax + b$ have multiple roots. This curve appears as bifurcation diagram in many different problems in singularity theory. For instance, it can be regarded as the bifurcation diagram of the zeros of the function x^3: it is formed by those points of the versal deformation of this function for which the 0 level set of a deformed function is singular. This bifurcation diagram has remarkable properties. For instance, its complement in $\mathbf{C}^2$ is an Eilenberg—MacLane space $K(\pi, 1)$: all homotopy groups of it are trivial, with the exception of the fundamental group (which is Artin's braid group on 3 strings, see Fig. 65).

Now consider a generic holomorphic vector field in the plane in which the semicubic parabola lives. As an example, consider the constant field $\partial/\partial b$ on the plane of cubic polynomials as above. At the cusp point of the semicubic parabola, the vector of this field is transversal to the tangent of the parabola. Any other vector field with this property can be reduced to this special field by a diffeomorphism preserving the semicubic parabola.

The semicubic parabola shares both the above properties—that of its complement being a space $K(\pi, 1)$ and that of admitting a rectification by a diffeomorphism of a generic field—with many other bifurcation diagrams. For instance, both properties hold for the swallowtail (Fig. 3), for bifurcation diagrams of zeros of simple functions on manifolds with boundary, and for bifurcation diagrams of simple projections.

We first consider functions on a manifold with boundary. Let $f : (\mathbf{C}^n, 0) \rightarrow (\mathbf{C}, 0)$ be a holomorphic function germ on the manifold $\mathbf{C}^n$, equipped with the hypersurface ('boundary') $\mathbf{C}^{n-1}$ defined by $x_1 = 0$ (x_k are the coordinates in $\mathbf{C}^n$). The point 0 is a *boundary singularity* of f if it is a critical point of the restriction of f to the boundary.

The *multiplicity μ of a boundary singularity* is the dimension, over $\mathbf{C}$, of the *boundary local algebra*

$$Q_f|_{x_1} = \mathbf{C}[[x_1, \ldots, x_n]]/(x_1 f_1, f_2, \ldots, f_n),$$

where $f_k = \partial f / \partial x_k$. The versal deformation can be chosen in the form:

$$F(x, \lambda) = f(x) + \lambda_1 g_1(x) + \cdots + \lambda_\mu g_\mu(x),$$

where the g_k are monomials whose classes in the quotient space $Q_f|_{x_1}$ form a basis (over $\mathbf{C}$). It is convenient to choose $g_\mu \equiv 1$. The space $\mathbf{C}^\mu$ with coordinates $(\lambda_1, \ldots, \lambda_\mu)$ is called the *base* of the versal deformation, while the space $\mathbf{C}^{\mu-1}$ with coordinates $(\lambda_1, \ldots, \lambda_{\mu-1})$ is called the base of the truncated versal deformation, or the *truncated base*.

Example. For the boundary singularity C_3 we choose

$$f = x_2^3 + x_1 x_2, \qquad F = f + \lambda_1 x_2^2 + \lambda_2 x_2 + \lambda_3.$$

Definition. A number α is a *critical value* of $F(\cdot, \lambda)$ if the variety $F(\cdot, \lambda) = \alpha$ is either singular or nontransversal to the boundary (i.e, to the hypersurface $x_1 = 0$).

Definition. The *bifurcation diagram of the zeros of a boundary singularity* is the set of those points in the base of the versal deformation for which 0 is a critical value of $F(\cdot, \lambda)$. This bifurcation diagram is denoted by Σ.

Example. The bifurcation diagram of the zeros of the simple singularity C_3 is formed by those points (a, b, c) for which the polynomial $x^3 + ax^2 + bx + c$ either has a multiple root or a root at 0. This diagram is shown in Fig. 90. It has 2 irreducible components. One of these (the surface corresponding to the multiple roots) is diffeomorphic to the cylinder over a semicubic parabola. The other component is the plane $c = 0$. It corresponds to the root at 0. The first component corresponds to the 0-level varieties $F(\cdot, \lambda) = 0$ that are not transversal to the boundary. The second component corresponds to the singular 0-level varieties (in this example, $\lambda_1 = a, \lambda_2 = b, \lambda_3 = c$).

The bifurcation diagrams of the zeros of simple boundary singularities coincide with the discriminants (the irregular orbit spaces) of the corresponding reflection groups (§ 4.2). These algebraic varieties have a well-defined tangent plane at the origin ($d\lambda_\mu = 0$ in our notation).

Theorem 1. *A holomorphic vector field that is transversal to the tangent space of the bifurcation diagram of the zeros of a simple boundary singularity is locally reducible to the constant vector field $\partial/\partial\lambda_\mu$ by a holomorphic diffeomorphism preserving the bifurcation diagram.*

This theorem, discovered in [98], was proved by O.V. Ljashko in [159], [160]. V.M. Zakaljukin [29] has extended it to certain nonsimple singularities.

Ljashko's proof is based on theorem 2 stated below.

Definition. The *bifurcation diagram of functions (for a boundary singularity)* is formed by those points $\overline{\lambda}$ in the truncated base for which the deformed function $F(\cdot, \overline{\lambda}, \lambda_\mu)$ has less than μ distinct critical values (note that the value of λ_μ is irrelevant, since $F(\cdot, \overline{\lambda}, \lambda_\mu) = F(\cdot, \overline{\lambda}, 0) + \lambda_\mu$).

Example. The bifurcation diagram of functions of the boundary singularity C_3 is formed by the straight line $b = 0$ and 2 parabolas tangent to it at the origin of the (a, b)-plane (Fig. 90).

The bifurcation diagram of functions of the boundary singularity C_4 is shown in Fig. 91. The (truncated) versal deformation is:

$$F = f + ax^3 + by^2 + cy, \qquad f = xy + x^4,$$

$$\text{boundary: } x = 0.$$

The bifurcation diagram is quasihomogeneous (invariant under the quasihomogeneous dilations $(a, b, c) \mapsto (ta, t^2b, t^3c)$). Hence only the intersection with the plane $a = \text{const}$ is shown.

The bifurcation diagram of a boundary singularity C_k consists of 4 hypersurfaces in the truncated base space. Indeed, it is formed by the caustic and the Maxwell set. The caustic has 2 components:

A^2 (the restriction to the boundary has a nonMorse critical point),

C_2 (a critical point falls on the boundary).

The Maxwell set has 2 components too:

$A^1 A_1$ (boundary and interior critical values coincide),

$2A^1$ (two boundary critical values coincide).

The caustic of C_4 is formed by those polynomials $y^4 + ay^3 + by^2 + cy$ that have either a nonMorse critical point (A^2) or a zero critical point (C_2). Hence it is diffeomorphic to the discriminant of C_3 (similarly, the caustic of C_{k+1} is diffeomorphic to the discriminant of C_k).

Fig. 91 shows that the Maxwell set of C_4 is diffeomorphic to its caustic. A priori this is not clear at all. This result extends Givental's theorem concerning 2 unfurled swallowtails, discussed at the end of § 4.4, to the case of C_k (or B_k). It would be interesting to understand the reason for this rather strange relation between caustics and Maxwell sets.

We return to the general theory of bifurcation diagrams of functions of boundary singularities. Generalising the Ljashko—Looijenga construction, we associate with any point $\overline{\lambda}$ in the truncated base space the value of the free term in the versal deformation for which the sum of the μ critical values of the functions $F(\cdot, \overline{\lambda}, \lambda_\mu)$ vanishes. Let us construct a polynomial whose roots are these critical values. We thus have constructed a map from the truncated base space to the space $\mathbf{C}^{\mu-1}$ of polynomials of degree μ with leading coefficient 1 and with vanishing subsequent coefficient.

Theorem 2 (see [160]). *The map above is proper and holomorphic, provided that the boundary singularity is simple. The complement of the bifurcation diagram of functions covers the manifold of polynomials without multiple roots.*

Corollary. *The complement of the bifurcation diagram of a simple boundary singularity is a space $K(\pi, 1)$, where π is a subgroup of finite index in Artin's braid group on μ strings.*

In order to deduce theorem 1 from theorem 2, we consider the μ moments when the particle whose motion is governed by our vector field passes through the discriminant. These moments define the straight line into which the orbit of the particle should be transformed by the required diffeomorphism, up to a choice from finitely many possibilities.

A deformation of the standard vector field into our field can be covered by a unique deformation of the identity diffeomorphism; this gives the required diffeomorphism preserving the discriminant (for details see [159], [160]).

Theorem 1 yields information about vector fields that are transversal to discriminants of ordinary (nonboundary) singularities. For instance, it implies that it is possible to straighten a generic vector field at the vertex of an ordinary swallowtail by a diffeomorphism preserving the swallowtail.

In order to reduce to normal form a vector field in a neighborhood of a vertex of the bifurcation diagram of the zeros of a function g, we apply the theorem to the function

$$f(x) = x_0 + g(x_1, \ldots, x_n)$$

on the manifold with boundary $x_0 = 0$. In this case the bifurcation diagram has only 1 component: it coincides with the bifurcation diagram of the restriction of f to the boundary, that is, to the bifurcation diagram of the zeros of g, since the level manifolds of the functions $F(\cdot, \lambda)$ are nonsingular.

Corollary. *The projection to a plane, along the c-axis, of the swallowtail formed by the polynomials $x^4 + ax^2 + bx + c$ that have multiple roots is stable: in a neighborhood of the origin, any nearby projection is equivalent to it.*

Of course, a similar corollary holds for all simple singularities of functions (on manifolds with or without boundary) and for the discriminants of all reflection groups (including unitary reflection groups).

The application of theorem 1 to the study of normal forms and perestroikas of various geometrical objects in space equipped with a bifurcation diagram (for instance, applications to the theory of projections of smooth hypersurfaces) are discussed in [98], where the theorem was first stated. Along with theorem 1, theorem 3, stated below, is very useful in the study of normal forms. In this theorem the bifurcation diagram of zeros is replaced by an arbitrary hypersurface V in $\mathbf{C}^n$ (singular or not).

Assume that the restriction of the fibration $p : \mathbf{C}^n \twoheadrightarrow \mathbf{C}^{n-1}$ to a (germ of a) hypersurface $V \subset \mathbf{C}^n$ is proper. Then locally V can be defined by an equation $R(x) = 0$, where R is a polynomial along any fiber of p and where the degree k of R is equal to the multiplicity of the intersection of V and the fiber at the origin. Let $D : \mathbf{C}^{n-1} \to \mathbf{C}$ be the discriminant of R. The variety on which the discriminant vanishes will be called the *bifurcation diagram of the projection* $V \hookrightarrow \mathbf{C}^n \twoheadrightarrow \mathbf{C}^{n-1}$.

Definition. A vector field v is said to *preserve a variety* $f = 0$ if the derivative of f along v belongs to the ideal generated by f.

Theorem 3 (see [159]). *Suppose that the dimension of the set of points in the base at which more than 2 points of the intersection of a hypersurface V and the fiber collapse (making the total number of distinct intersections of the fiber and V smaller than $n - 1$) is strictly less than the dimension of the bifurcation diagram.*

Then every holomorphic vector field germ on the base that is tangent to the bifurcation diagram admits a lifting to a holomorphic vector field germ in the space of the fibration $\mathbf{C}^n \twoheadrightarrow \mathbf{C}^{n-1}$, preserving the hypersurface V.

Corollary. *A holomorphic vector field preserving the bifurcation diagram of functions of a simple singularity admits a lifting to a holomorphic vector field that is tangent to the bifurcation diagram of zeros (to the discriminant).*

Example. A field that is tangent to the union of a semicubic parabola and the tangent line to the latter at the cusp, admits a lifting to the swallowtail whose cuspidal edge and line of selfintersection project onto the semicubic parabola and this tangent line.

These infinitesimal results have global variants, in which the vector fields are replaced by finite diffeomorphisms. In order to lift a diffeomorphism that preserves a bifurcation diagram, it is useful to know the group π_0 of connected components of the space of biholomorphic diffeomorphisms preserving the diagram.

For discriminants of reflection groups the answer was found by O.V. Ljashko [159],

[160]:

Theorem 4. *The group π_0 coincides with the quotient of the group of automorphisms of the reflection group, preserving the set of reflections, by the invariant subgroup of inner automorphisms. This group π_0 contains 2 elements for the cases D_{2k}, B_2, F_4, G_2, $I_2(2k)$, and is isomorphic to the group of permutations of 3 elements for the case D_4. In all other cases it is trivial.*

We now return to the simple projections of complete intersections. The above theorems of Ljashko were extended to this case by V.V. Gorjunov. His list of simple projections (§ 6.2) includes the list of simple boundary singularities. But even in this case Gorjunov's theorems provide other information than Ljashko's theorems do: starting from the same boundary singularities, Gorjunov obtains *other* Eilenberg—MacLane spaces and straightens *other* vector fields.

Example. Consider once more the space of cubic polynomials $x^3 + ax^2 + bx + c$ having multiple roots. It is diffeomorphic to the cylinder over a semicubic parabola, and it forms a part of the bifurcation diagram of zeros of C_3. This time we project it not along the c-axis (as in Ljashko's theorem and in Fig. 90), but along the b-axis (Fig. 92). In this case the vertical plane, indicated by dots in Fig. 92, does not belong to the projected manifold.

The *apparent contour* of the projection on the (a, c)-plane consists of 2 smooth curves, A^2 and C_2, which are tangent of order 3 at the origin. In this example Gorjunov's theorems are as follows [131], [161].

Theorem 5. *The complement of the apparent contour is the Eilenberg—MacLane space $K(\pi, 1)$, where π is a subgroup of index 8 in Artin's braid group on 3 strings.*

Theorem 6. *The projection along the b-axis of the above semicubic cylinder over the semicubic parabola is stable: any nearby (possibly nonlinear) holomorphic projection of the cylinder is locally biholomorphically equivalent to the projection $(a, b, c) \mapsto (a, c)$.*

Theorem 7. *The germ at the origin of the vector field $\partial/\partial b$ is stable with respect to the above semicubic cylinder: any nearby holomorphic vector field is reducible (at some point near the origin) to the form $\partial/\partial b$ by a local biholomorphic diffeomorphism of the 3-space preserving the semicubic cylinder.*

Theorem 8. *Any generic holomorphic vector field is, at any point of the cuspidal edge of the semicubic cylinder $u^2 = v^3$, reducible (by a diffeomorphism in the 3-space with coordinates (u, v, w) preserving the cylinder) to one of the 2 normal forms $\partial/\partial u$ (for generic points), $\partial/\partial v + w\partial/\partial u$ (for singular points).*

In order to formulate the general results for an arbitrary projection onto a line, consider the projection to the u-axis of the complete intersection V_0 defined by the equations $f(x, u) = 0$. A deformation of this projection is a family of projections $(x, u) \mapsto u$ of the complete intersections V_λ defined by equations $F(x, u, \lambda) = 0$, where $F(x, u, 0) \equiv f(x, u)$. Assume that F is a miniversal deformation of a simple germ of a holomorphic projection. That is, assume that the number n of parameters λ is as small as possible (for the definition of versal deformations see § 6.2).

Example. For the simple projection germ C_3 the complete intersection V_0 is the planar curve $f(x, u) = 0$ where $f = x^3 + ux$. A miniversal deformation is a 2-parameter family of curves $F(x, u, \lambda) = 0$ where $F = f + \lambda_1 x^2 + \lambda_2$.

The deformation of a projection is associated with the 'total complete intersection'

$$W = \{(x, u, \lambda) : F(x, u, \lambda) = 0\}.$$

If the deformation is versal, W is a smooth manifold at the origin.

Definition. The *bifurcation diagram of zeros of a projection* is the germ at the origin of the set of critical values of the restriction of the map $(x, u, \lambda) \mapsto (u, \lambda)$ to the (germ of the) manifold W.

Example. The bifurcation diagram of zeros of the projection C_3 is formed by those $(u.\lambda)$ for which the polynomial $x^3 + \lambda_1 x^2 + ux + \lambda_2$ has a multiple root (in Fig. 92, $(\lambda_1, u, \lambda_2)$ are denoted by (a, b, c)).

Now we consider the natural projection to the parameter space $\{\lambda\}$ along the u-axis of the bifurcation diagram of zeros of a projection.

Definition. The *bifurcation diagram of a projection* is the union of the projection to the parameter space $\{\lambda\}$ of the set of singular points of the bifurcation diagram of zeros and the set of critical values of this projection on the smooth part of this bifurcation diagram of zeros.

Example 1. The projection of the smooth part of the bifurcation diagram of zeros of C_3 to the $\{\lambda\}$-plane (with coordinates (a, c), Fig. 92) has the a-axis as set of critical values. The projection of the set of singular points of the bifurcation diagram of zeros is a cubic parabola. Together they form the bifurcation diagram of the projection C_3. It is not diffeomorphic to the bifurcation diagram of the function C_3 (which consists of 3 components that are tangent of order 2).

Example 2. The projection of the swallowtail to the plane, along a direction transversal to the tangent plane at the vertex, does not have critical values on the smooth part of the swallowtail. The set of singular points of the swallowtail consists of 2 components: the cuspidal edge and the line of selfintersection. Their projections to the plane form the bifurcation diagram of the projection A_3 (it coincides with the bifurcation diagram of the function A_3). Here $f = u + x^4$, $F = f + \lambda_1 x^2 + \lambda_2 x$.

In the general case the bifurcation diagram of a projection consists of 3 parts: the projection of the cuspidal edge of the bifurcation diagram of zeros, the projection of the set of selfintersections, and the set of critical values of the projection of the regular part.

The bifurcation diagram of a projection can also be defined as the set of those parameter values λ in the miniversal deformation for which either the variety $V_\lambda = \{(x,u) : F(x,\lambda,u) = 0\}$ is singular, or the restriction of u to this variety is not a Morse function.

Theorem 9 (see [131], [133]). *The (germ of the) complement of the bifurcation diagram of a simple projection from a complete intersection of positive dimension onto the line is an Eilenberg—MacLane space $K(\pi, 1)$, where π is a subgroup of finite index in Artin's braid group on $n + 1$ strings (here $n = \dim\{\lambda\}$).*

Theorem 10. *The germ at the origin of the vector field $\partial/\partial u$ is stable with respect to the bifurcation diagram of zeros of a simple projection: any nearby vector field is reducible to this normal form in some nearby point by a biholomorphic diffeomorphism preserving the bifurcation diagram of zeros.*

A similar result holds for nonsimple projections, provided that there exist a quasi-homogeneous versal deformation of them.

Example. Consider in $\mathbf{C}^n$ the hypersurface Σ of polynomials $x^n + a_1 x^{n-1} + \cdots + a_n$ having multiple roots. This hypersurface is diffeomorphic to the cartesian product of a swallowtail in $\mathbf{C}^{n-1}$ and the line $\mathbf{C}$. The product of the vertex of the swallowtail and $\mathbf{C}$ defines the most singular line on Σ. The theorems of Ljashko and Gorjunov provide the normal forms of generic holomorphic vector fields at all points of this singular line.

Indeed, at generic points of this line the vector field is transversal to the tangent hyperplane of Σ. In a neighborhood of such a point the vector field is reducible to the normal form $\partial/\partial a_n$ by a diffeomorphism preserving Σ (theorem 1). At certain isolated points of the most singular line the vector field belongs to the tangent hyperplane, but otherwise it is generic (provided we started with a generic vector field). In a neighborhood of such an isolated point the field is reducible to the normal form $\partial/\partial a_{n-1}$ (theorem 10).

Remark. The theorems above have many generalisations. For instance, Gorjunov has extended them to the case of simple projections from complete intersections with boundary to the line (they are stably equivalent to the simple projections of hypersurfaces with boundary), and to the case of the simple linear singularities introduced by Siersma [162].

A nonisolated critical point of a function is called *linear* if the set of critical points forms a smooth line, and if the singularity of the restriction of the function to the generic hyperplane transversal to the line of critical points is nondegenerate (of Morse type).

Gorjunov has found that the simple linear singularities are classified by the reflection groups. The hierarchy of nonisolated singularities reflects the hierarchy of the sequence of singularities. For instance, the whole sequence D_k corresponds to the single nonisolated singularity D_∞. Thus, a Coxeter type corresponds to a sequence of singularities. For instance, in the notation of [28],

$$A_n \sim J_{n+1,\infty}, \qquad B_n \sim T_{2n+2,\infty,\infty};$$

$$C_n \sim Z_{n-2,\infty}, \qquad F_4 \sim W_{1,\infty}.$$

For these singularities Gorjunov has defined bifurcation diagrams and has proved the $K(\pi,1)$-property [80].

It is interesting to note that for simple objects the $K(\pi,1)$-property holds only for 'natural' theories, and that in this sense projections are more natural objects than complete intersections (the state of the theory of distinguished bases, discussed in § 6.2, leads to the same conclusion).

Example. Consider the bifurcation diagram of zeros of the 0-dimensional complete intersection

$$x^2 + u^2 = 0, \qquad xy = 0$$

(denoted by $I_{2,2}$ in Giusti's list [154] of simple singularities of complete intersections). Knörrer [129] has proved that the complement in $\mathbf{C}^4$ of this diagram has nontrivial group π_2, hence is not a space $K(\pi,1)$.

However, Gorjunov [130] has observed that the complement of the bifurcation diagram of the simple *projection* to the u-axis from the same complete intersection *is indeed* a space $K(\pi,1)$.

The bifurcation diagram of the complete intersection is

$$\{\lambda \in \mathbf{C}^4 : \text{ the variety}$$

$$u^2 + \lambda_1 u + \lambda_2 + \lambda_3 x + x^2 = 0, \ ux = \lambda_4 \text{ is singular}\}.$$

The bifurcation diagram of the projection $(x,u) \mapsto u$ is the union of the bifurcation diagram of the complete intersection and the hyperplane $\lambda_4 = 0$ (above this hyperplane the number of distinct values of u on the complete intersection is less than above the complement).

Thus, projections are better objects than complete intersections (however, there still exist simple projections of complete intersections for which the $K(\pi,1)$-conjecture has not been proved; namely, $X_{k,l}$, $3 \le k < l$, V_6, V_7, $\Gamma^{a,b}_{a+b+3}$).

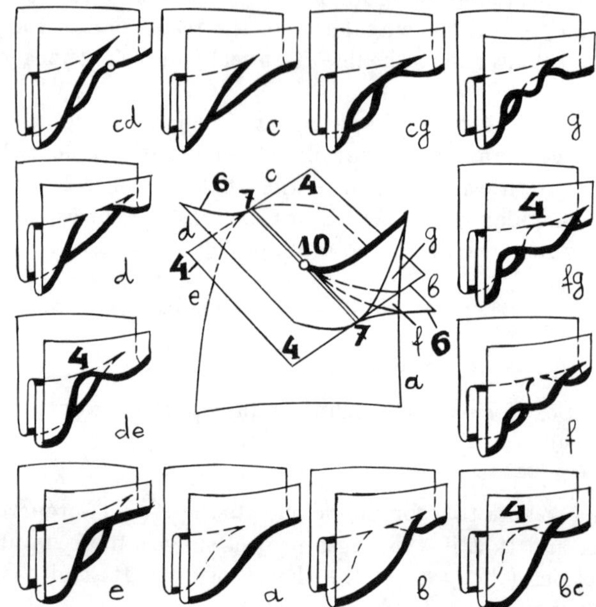

Figure 81: The bifurcations of the projection $x^4 + x^2y + xy^3$

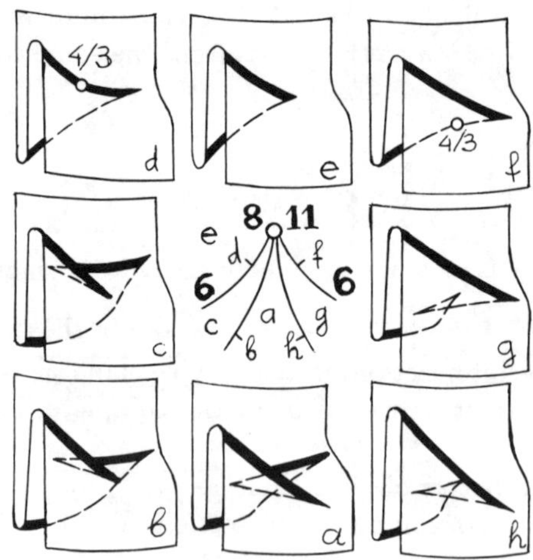

Figure 82: The bifurcations of the projection $x^5 + xy + axy^3$

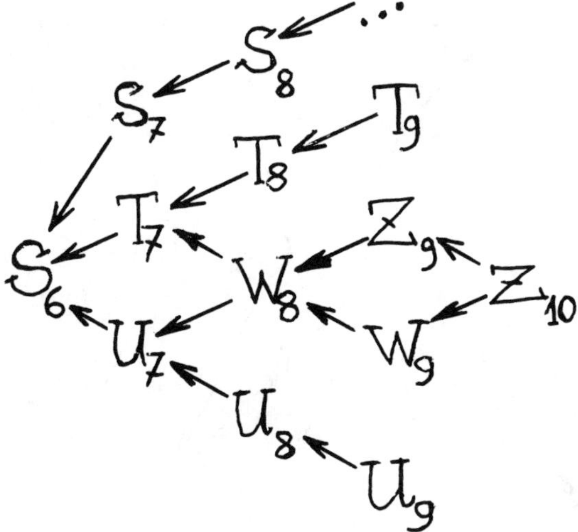

Figure 83: The hierarchy of simple space curves $f_{1,2}(x, y, z) = 0$

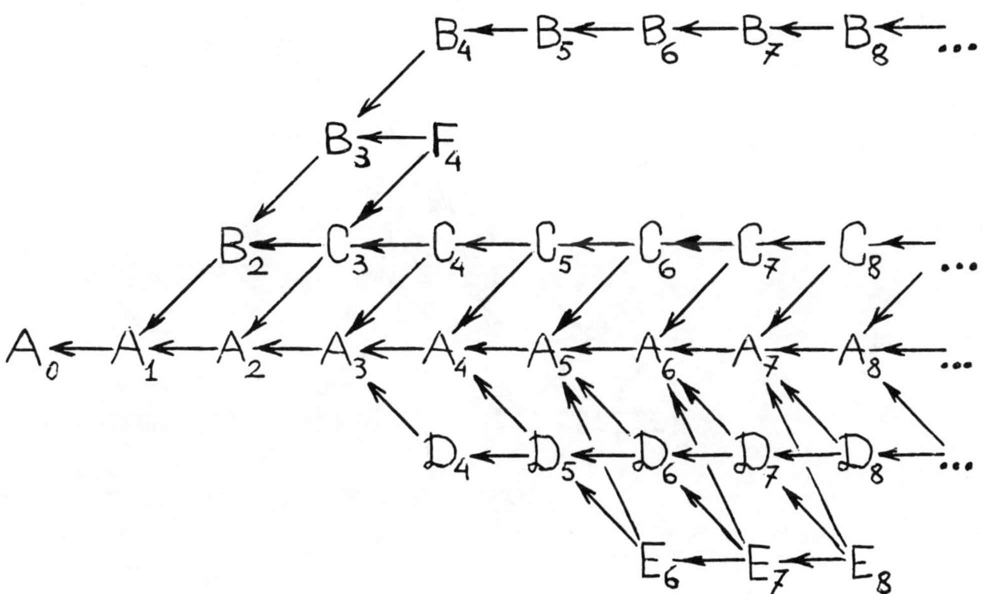

Figure 84: The hierarchy of simple projections of hypersurfaces

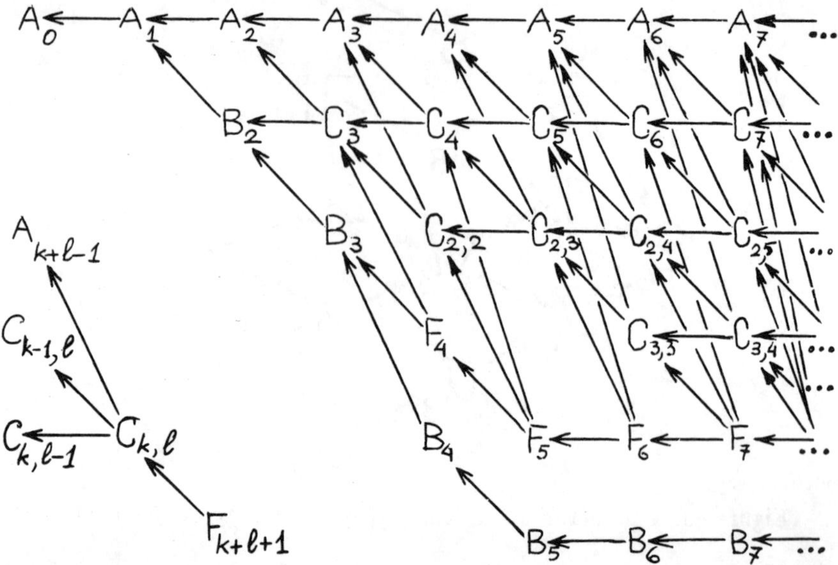

Figure 85: The hierarchy of simple projections of spatial curves

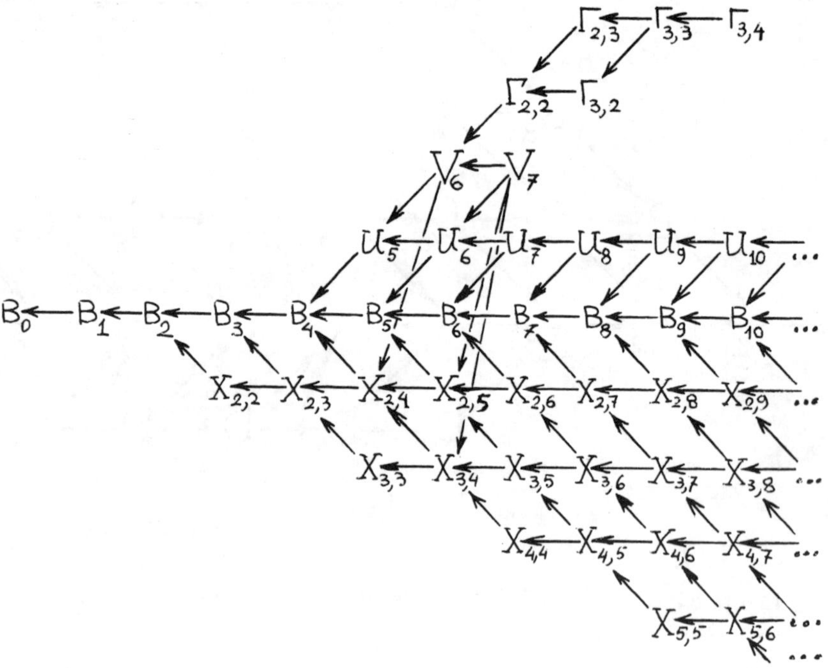

Figure 86: The hierarchy of simple projections of multiple points

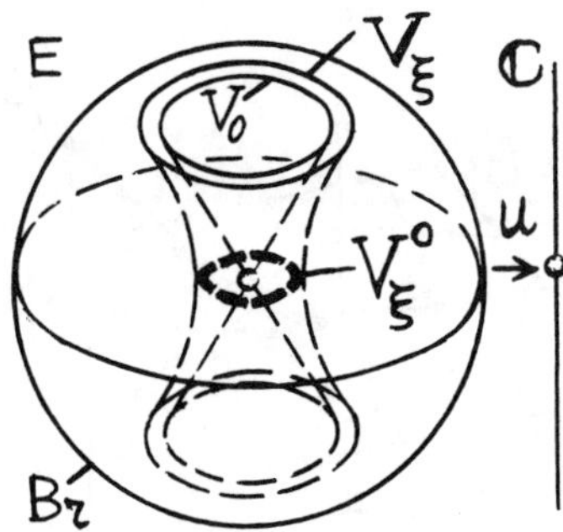

Figure 87: A Milnor fiber of a projection

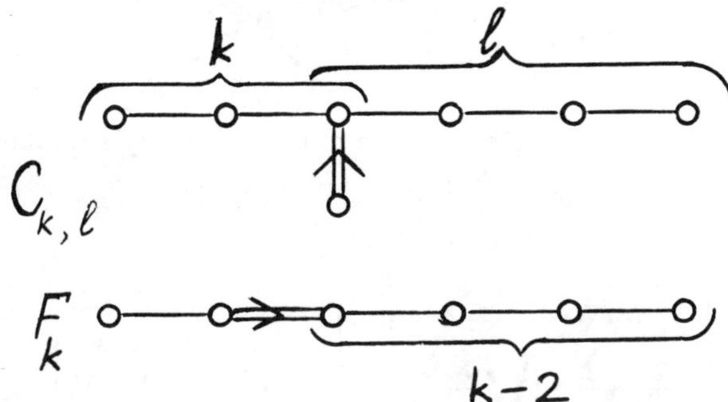

Figure 88: The Dynkin diagrams $C_{k,l}$ and F_k

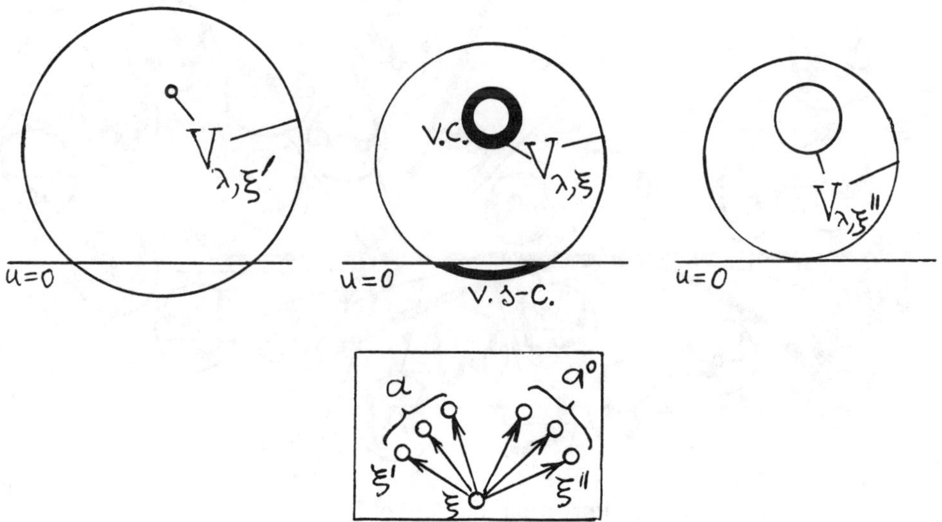

Figure 89: Vanishing cycles and semicycles

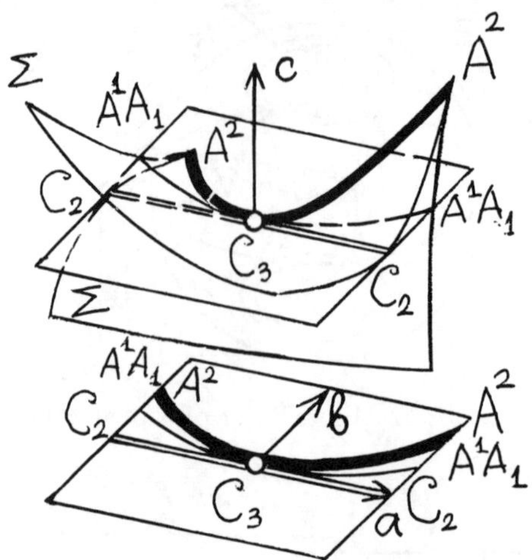

Figure 90: The bifurcation diagram of the boundary singularity C_3

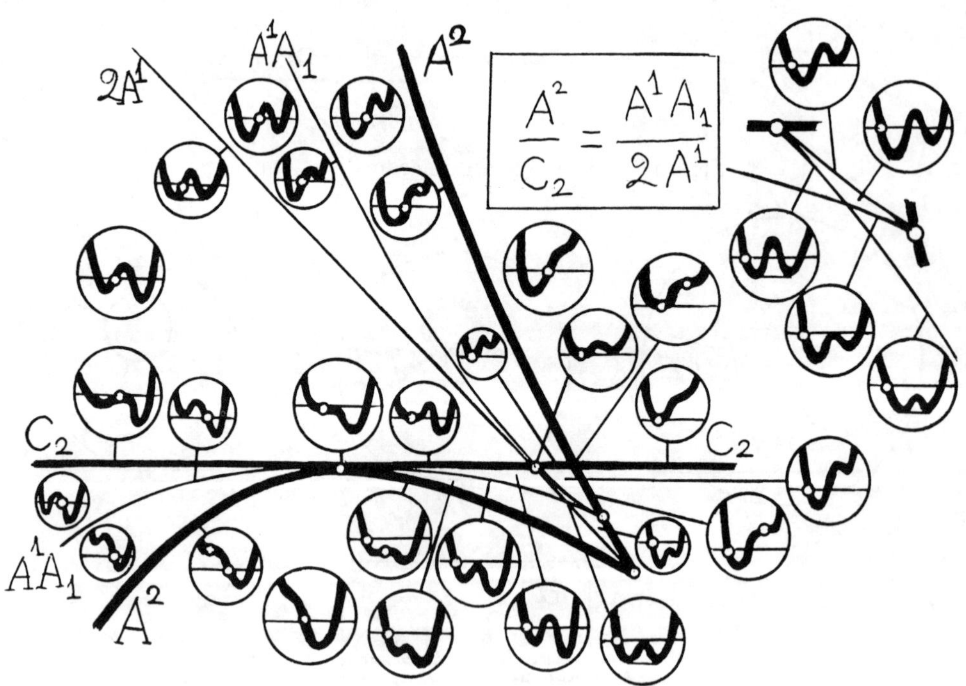

Figure 91: The bifurcation diagram of functions for C_4

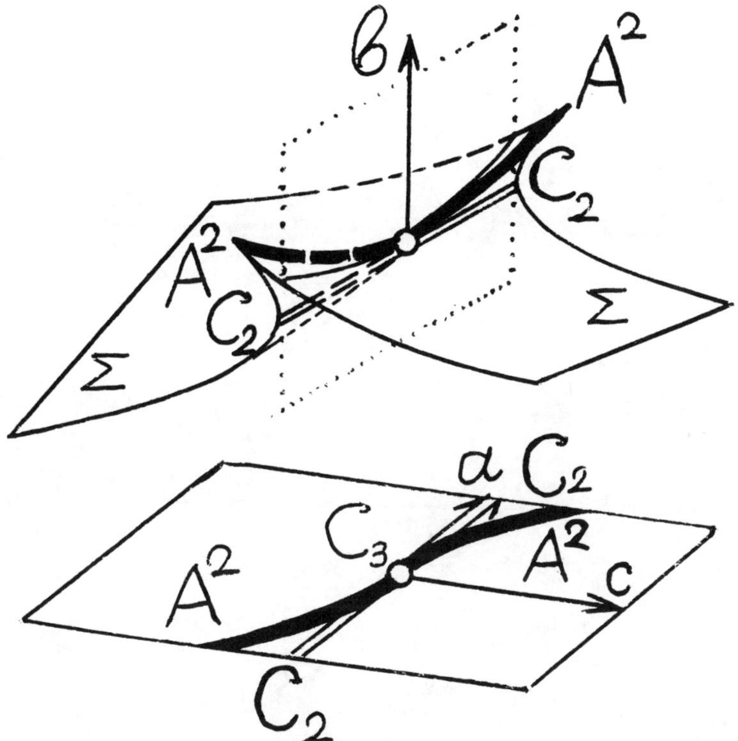

Figure 92: The bifurcation diagram of the projection C_3

Chapter 7

Obstacle problem

Consider in a euclidean space an obstacle bounded by a smooth generic surface. The obstacle problem is to find the singularities of the distance function from a point in the space to a given initial set (say, a point) along paths that do not intersect with the inside of the obstacle.

For an obstacle bounded by a planar curve the level lines of the distance function are the evolvents (= involutes) of the curve (see Fig. 93). Hence these level lines have semicubical cusps at the boundary points of the obstacle, provided that the boundary is convex (provided that the curvature of the boundary does not vanish).

At a generic inflection point of the boundary the evolvent has a singularity E_8 of order 5/3 (this is shown by a simple computation). The Legendre variety corresponding to this front (in the space of contact elements on the plane) is itself singular (it has a semicubical cusp).

Fronts of variational problems on manifolds without boundary (for instance, equidistant hypersurfaces in smooth manifolds) may also have singularities. These singularities, however, are singularities of the projection: the projected Legendre manifold is smooth. For variational problems on manifolds with boundary (and for the more general variational problems with one-sided constraints) the Legendre varieties formed by the contact elements that are tangent to the fronts, are singular themselves.

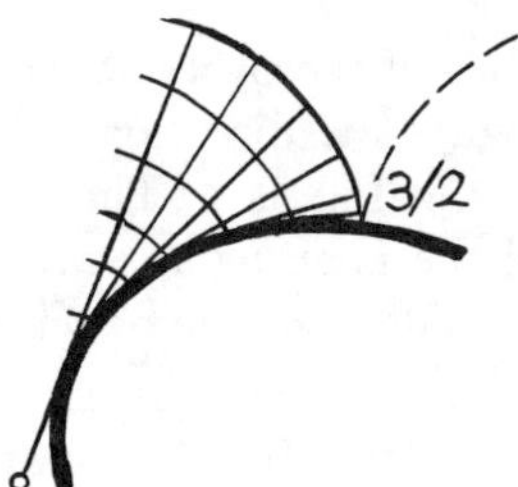

Figure 93: Evolvents as wave fronts

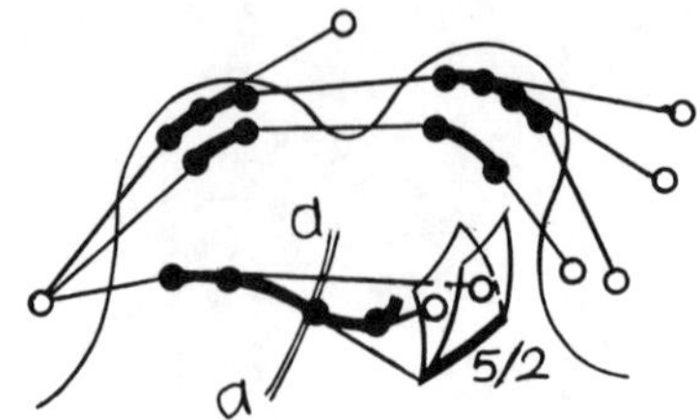

Figure 94: The extremals of the obstacle problem

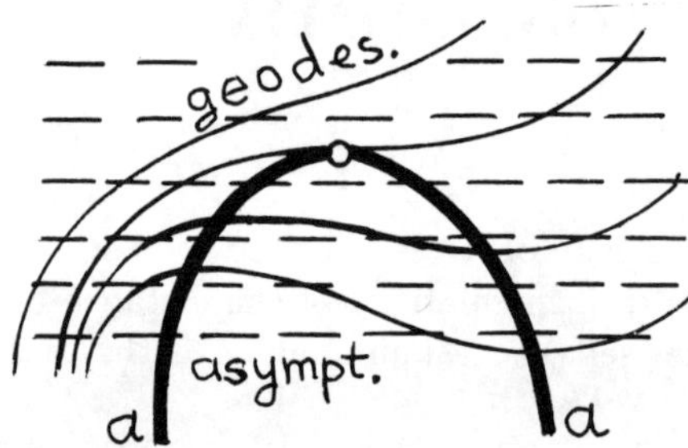

Figure 95: The field of asymptotic directions, and the pencil of geodesics

Singular Lagrangian varieties arise from the same source. The extremals joining a point in a manifold with other points form a Lagrangian variety in the space of extremals. This Lagrangian variety is smooth if the initial manifold has no boundary, but it may be singular otherwise.

Consider, for instance, the shortest paths from a fixed point in euclidean 3-space to arbitrary points outside an obstacle (Fig. 94). The initial part of such a path is a straight line joining the initial point with the point of tangency of the line with the boundary surface of the obstacle. The next part is a segment of a geodesic of this boundary surface. Thus, a pencil (a 1-parameter family) of geodesics on the surface of the obstacle arises. The next part of the path lies on the straight line leaving the surface along the tangent to a geodesic of the pencil at a certain point of the surface, etc.

At certain points of the surface of the obstacle, the direction of a geodesic in the pencil will coincide with an asymptotic direction of the surface. Generically, such points form a smooth line (line a in Fig. 95). The rays leaving the surface at these points form the cuspidal edge of the Lagrangian variety of outgoing rays. At certain isolated points of the line a the geodesic of the pencil is tangent to this line.

Generically, the ray leaving the surface of the obstacle at such a point on a is the vertex of the unfurled swallowtail singularity of the variety of outgoing rays (Fig. 10). In this way the unfurled swallowtail enters modern singularity theory (see [23]).

The Lagrangian and Legendre varieties describing the solutions of variational problems with one-sided constraints always have singularities of this type. Although the generic obstacle problem has not even been solved in euclidean 3-space, in this chapter we will describe the normal forms of the most important Lagrangian and Legendre varieties occurring in these problems, including the normal forms of the corresponding families of rays in the space of extremals.

These normal forms come from the theory of invariants of binary forms. The quite unexpected relation between the theory of binary forms and the calculus of variations was discovered in a long series of papers ([23]–[25], [72], [11], [12], [4], [5], [8], [143], [144]) as a result of attempts to understand the deep reasons for the coincidences of bifurcation diagrams in various theories, for the miraculous cancellation of many terms in long formulas, and for the strange universality of the unfurled swallowtail.

From another side, the graph of the time function in the planar obstacle problem is locally diffeomorphic to the variety of nonregular orbits of the icosahedron symmetry group (the group H_3 in Coxeter's classification of symmetry groups). In the spatial (3-dimensional) obstacle problem this very surface occurs as the singularity of the front (at a point of tangency of an asymptotic ray with the surface of the obstacle).

The discriminant of the hypericosahedron symmetry group H_4 occurs as the singularity of the graph of the distance function at a point of the asymptotic ray leaving the surface at a parabolic point of it. Hence we begin our analysis of the obstacle problem with a general discussion of the geometry of asymptotic tangent lines of surfaces.

7.1 Asymptotic rays in symplectic geometry

A generic plane curve does not have tangent lines of order of tangency exceeding 2. A generic surface in euclidean 3-space does not have tangent lines of order of tangency exceeding 4. Tangent lines of order of tangency higher than 1 (asymptotic lines) exist in the whole domain of hyperbolicity, those with order of tangency higher than 2 exist on a curve, while those with order of tangency 4 exist at isolated points of the latter line (see Fig. 77).

The hierarchy of asymptotic tangent lines becomes more transparent when formulated in the language of symplectic (or contact) geometry. This reformulation extends the theory of asymptotic tangent lines to a wide class of new situations; for instance, to the case of submanifolds of a riemannian space or to general variational problems with one-sided constraints.

Consider a hypersurface ∂M in a riemannian manifold with boundary M. The points of the phase space T^*M will be called *vectors*. The riemannian metric determines in the phase space the hypersurface of vectors of length 1. The surface ∂M determines in the phase space the hypersurface of vectors based at points of ∂M.

A large part of the exterior geometry of ∂M in M may be formulated in terms of the symplectic geometry of pairs of hypersurfaces in symplectic manifolds (this was observed by R. Melrose [18] for the Birkhoff billiard problem when M is a domain bounded by ∂M). Neither the cotangent bundle structure of the ambient symplectic phase space nor the origin of the hypersurfaces do play a role: we can start from any

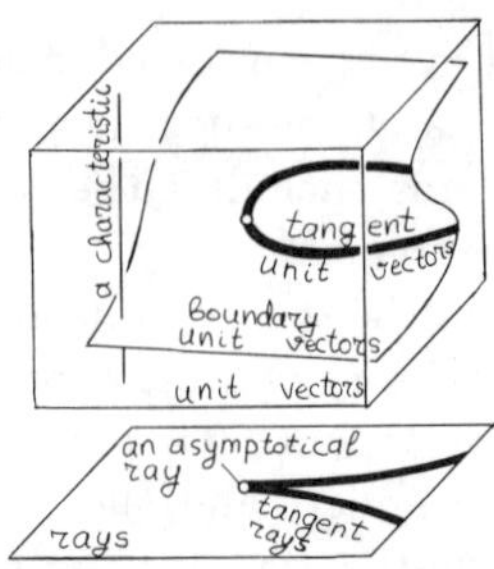

Figure 96: The asymptotic unit vectors as singular points of the canonical projection

pair of hypersurfaces in any symplectic manifold. In this way we can use geometrical intuition, stemming from experiments with surfaces in the ordinary euclidean space, in the general setting of variational problems with one-sided constraints.

A. Geometry of pairs of hypersurfaces of symplectic manifolds.

Consider two generic smooth hypersurfaces in a symplectic manifold. We will call one of them the 'surface of unit vectors' and the other the 'surface of boundary vectors'. We assume that they intersect transversally along the submanifold of 'unit boundary vectors' (this submanifold has codimension 2 in the initial symplectic manifold). A hypersurface in a symplectic manifold locally fibers into characteristics (integral curves of the field of skeworthogonal complements of the tangent hyperplanes). The characteristics of the 'surface of unit vectors' will be called 'rays' (if this surface is transversally oriented, the 'rays' have a natural orientation).

The space of 'rays' is even-dimensional, and inherits from the ambient space a natural symplectic structure (see § 1.1). The 'manifold of unit vectors' is fibered over the space of 'rays', and includes as a hypersurface the 'submanifold of unit boundary vectors'. The pair formed by the inclusion and the projection defines a *canonical projection* from the space of 'unit boundary vectors' to the space of 'rays': send each 'unit boundary vector' to the 'ray issuing from this vector'. The projection is a map between 2 spaces of equal dimension.

The singularities of the canonical projection constructed from a pair of generic hypersurfaces are the standard Whitney singularities. This means that locally they are equivalent to the singularities A_k (i.e. those of projections of a hypersurface

$$\{(x, \lambda) : x^{k+1} + \lambda_1 x^{k-1} + \cdots + \lambda_k = 0\}$$

in the total space of a fibration $(x, \lambda) \mapsto \lambda$ to the base space $\{\lambda\}$).

Example. $k = 1$ corresponds to the usual fold singularity: the parabola $x^2 + \lambda_1 = 0$ is projected to the λ_1-axis.

Definition. A 'unit boundary vector' is called '*tangent*' if it is a singular point of the canonical projection. It is called '*asymptotic*' if the corresponding singularity of

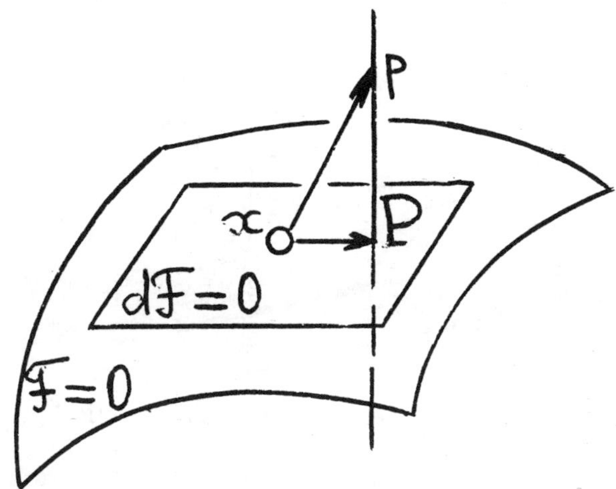

Figure 97: A characteristic of the surface of boundary vectors

the projection is more complicated than the fold $(k > 1)$. For $k = 2$ it is called *biasymptotic*, etc. (Fig. 96).

Example. For the surface $F(x) = 0$ in the ordinary euclidean 3-space, and for the surface of unit vectors $\{(x, p) : p^2 = 1\}$, the definitions above agree with the usual definitions in differential geometry.

Along with the above-introduced canonical projection we can define the *second canonical projection* by the same construction, in which the hypersurfaces are interchanged.

Returning to the above example, consider the momentum p as a vector in euclidean space. A characteristic of the surface of boundary vectors consists of the vectors, based at the same point, whose end points belong to a straight line that is orthogonal to the tangent plane of the boundary at the base point (Fig. 97).

With this in mind, in the abstract setting we will call the (space of) characteristics of the 'hypersurface of boundary vectors' the (bundle of) '(co)tangent vectors of the boundary'.

The second canonical projection sends the manifold of 'unit boundary vectors' to the manifold of '(co)tangent vectors of the boundary'. In the example this projection sends a unit boundary vector to its orthogonal projection P to the tangent plane (Fig. 97). Hence in this example the only singularity of the second canonical projection is the fold (A_1) along the manifold of unit vectors tangent to the boundary (Fig. 98). Of course, the same is true for a riemannian manifold.

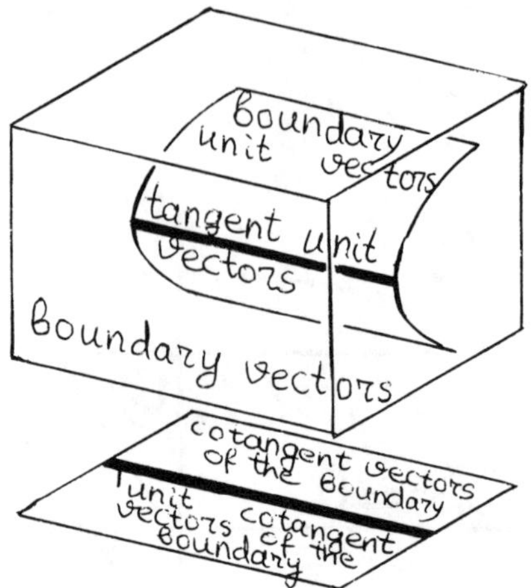

Figure 98: The second canonical projection

The meaning of the second canonical projection becomes more transparent if we regard the momentum p as a cotangent vector of the configuration space, that is, as a 1-form on the tangent space at the point x of the configuration space. The characteristic of the manifold of boundary vectors is the straight line formed by the 1-forms $p + \mathbf{R}\, dF$. This line is unambiguously defined by the restriction of the form p to the tangent space $dF = 0$ of the boundary. Hence, *the second canonical projection sends a 1-form to its restriction to the boundary.*

B. The normal forms.

If the initial surface in euclidean space is quadratically strictly convex (i.e. the second quadratic form is nondegenerate), then both the first and the second canonical projection have fold singularities (A_1) on the manifold of 'unit boundary vectors'. Assume that a pair of hypersurfaces in a symplectic space generates two fold singularities of the canonical projections. A fold map (locally) defines an *involution* of the initial manifold in a neighborhood of its hypersurface of critical points—it permutes the two pre-images of the points covered by the image of the map. In our case the two canonical projections have fold singularities. Hence we obtain *two* involutions on the manifold of 'unit boundary vectors'; they both fix the hypersurface of 'unit vectors tangent to the boundary'. These two involutions, defined by the first and second canonical projections, will be called the first and second *Melrose involutions*.

Example. For a planar convex curve the first involution sends a unit vector based at a point of the curve to a unit vector in the same direction, but based at the other point of intersection of the ray determined by the vector and the curve (Fig. 99). The second involution sends it to a vector based at the same point, but lying symmetrically to the initial vector with respect to the tangent line.

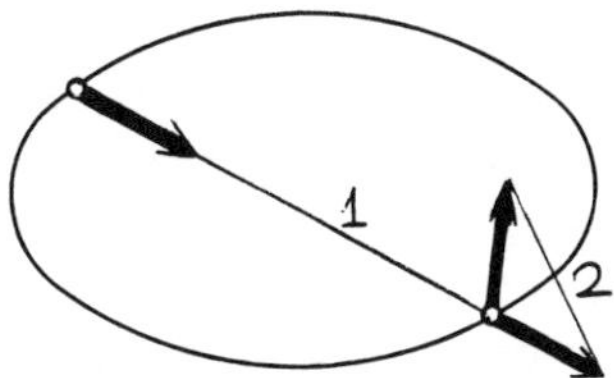

Figure 99: The Birkhoff billiard transformation as the product of two Melrose involutions

The product of two Melrose involutions is the *Birkhoff billiard transformation.*

In suitable Darboux coordinates ($\omega = dp_0 \wedge dq_0 + \cdots$), a pair of hypersurfaces is reducible to the local C^∞ normal form

$$q_0 = 0, \qquad q_0 = p_0^2 + p_1,$$

provided that the singularities of both canonical projections are folds (Melrose [18]). The objects constructed by us (for instance, Melrose involutions) are hence also reducible to normal form.

It is interesting to note that reduction to the above normal form is possible in the C^∞ situation and at the level of formal series, but not in the analytic or holomorphic situation. The pairs of involutions encountered in practice in the analytic or holomorphic situation usually have simple formal or C^∞ classifications and functional moduli spaces. In the problem of classifying pairs of hypersurfaces of a symplectic manifold, the functional moduli spaces have not yet been studied, as far as I know (this is also true for the local analytic classification of the corresponding pairs of symplectic involutions and their products).

The 'global' problem of classifying pairs of involutions along a complete closed fixed submanifold is a hopeless task, even at the topological level. Indeed, in the simplest case when this manifold is a circle, the product of the corresponding involutions is a symplectic map of an annulus, fixing the circle. The topological classification of such maps includes most of the difficulties inherent in nonintegrable problems in Hamiltonian dynamics (see [93]).

If the singularity of a canonical projection is more complicated than a fold, then there is no simple normal form for the pair of hypersurfaces (see [163]). However, for the next two singularities, A_2 and A_3 (for instance, for the usual asymptotic and biasymptotic rays of surfaces in euclidean space), it is still possible to reduce to normal form (at least at the level of formal series) the pair formed by the first hypersurface and its intersection with the second.

Theorem 1. *In a neighborhood of a point at which the canonical projection has a singularity A_k, a codimension-1 manifold of the hypersurface $q_0 = 0$ is reducible to the normal form $q_0 = 0 = F$, where*

$$F = \qquad p_0^2 + p_1 \qquad \text{for } k = 1,$$
$$F = \quad p_0^3 + p_1 p_0 + q_1 \qquad \text{for } k = 2,$$

$$F = \quad p_0^4 + p_1 p_0^2 + q_2 p_0 + p_2 \quad for\ k = 3.$$

Here suitable Darboux coordinates ($\omega = dp_0 \wedge dq_0 + \cdots + dp_n \wedge dq_n$) are used.

This result (which was proven at the level of formal series in [23]) presumably also holds in the C^∞ situation; however, a proof of this has never been published, as far as I know.

The theorem implies the possibility of reduction to simple normal form of objects derived from the pair formed by the first hypersurface and its intersection with the second; for instance, for the first canonical projection (which becomes the map of 'forgetting p_0').

The critical values of the first canonical projection form the variety of 'tangent rays', which has a semicubical cuspidal edge in the case of A_2, and a swallowtail singularity (times a smooth manifold) in the case of A_3. This variety of 'tangent rays' is a hypersurface in the symplectic space of rays.

The reduction of the pair (hypersurface, subhypersurface) to symplectic normal form implies a reduction to symplectic normal form of the variety of 'tangent rays'.

Thus, the results above imply local normal forms of symplectic structures in spaces containing as a hypersurface a variety that is diffeomorphic to the product of the semicubic parabola or a swallowtail surface and a smooth manifold:

Theorem 2. *A generic symplectic structure is formally reducible to the form*

$$dA \wedge dB + dC \wedge dD + \cdots$$

in a neighborhood of the point 0 of the hypersurface $A^3 = B^2$ (here, generic = non-degenerate on the space $A = B = 0$), or to the form

$$dA \wedge dD + dC \wedge dB + dE \wedge dF \cdots$$

in a neighborhood of the point 0 of the hypersurface defined by the condition:

$$x^4 + Ax^2 + Bx + C \quad has\ a\ multiple\ root$$

(here, generic = nondegenerate on the space $B = C = 0$).

The proof can be found in [23] (the assertions are probably true at least in the C^∞ situation).

These results can be interpreted as reduction to normal form, using symplectic diffeomorphisms, of products of the semicubic parabola or the swallowtail surface and a linear space.

A similar formal classification is impossible for $k > 3$, since moduli arise.

The contact versions of the above classifications are due to E.E. Landis [164], [165] (see also Melrose [163]):

$$\alpha = dZ - B\,dA - \cdots$$

in a neighborhood of the cuspidal edge of the cylinder $A^3 = B^2$;

$$\alpha = dZ - D\,dA - C\,dB - E\,dF - \cdots$$

in a neighborhood of the edge of the cylinder over the swallowtail surface.

C. Geodesics.

The set of critical values of a canonical projection is a hypersurface in a symplectic space. Its characteristics are called '*geodesics*'.

Example. Consider the pair of hypersurfaces in the phase space corresponding to a surface in euclidean space. The critical values of the first canonical projection are the tangent rays. In this case the 'geodesics' are the curves in the space of rays formed by the tangent rays of a geodesic line of the initial surface.

The critical values of the second canonical projection are the unit vectors that are (co)tangent to the surface. In this case the 'geodesics' are the orbits of the usual geodesic flow on the initial surface.

As in this example, we define two sets of 'geodesic lines': one is defined by the first, the other by the second canonical projection.

The third family consists of the characteristics on the manifold of critical points (which is the same for both projections). This manifold has codimension 3 in the ambient symplectic manifold. The *characteristic direction* is defined as the kernel of the restriction of the symplectic structure (it is skeworthogonal to the whole tangent space of the manifold of critical points).

The characteristics of the set of critical points will be called the '*upper geodesics*'. The canonical projections send upper geodesics to lower geodesics (at the fold points at which the latter are defined), see [166].

The above normal form classification of the first canonical projection implies the following normal forms of the first family of 'lower geodesics' and 'upper geodesics' [23]:

Theorem 3. *In the coordinates of theorem 1 above, the equations of the 'upper geodesics' are:*

$$q_0 \qquad\quad = p_0 = 0,\; q_1 = \xi \qquad\qquad \text{for } k = 1,$$
$$q_0 \;= 0,\; p_0 = \xi,\; p_1 = -3\xi^2,\; q_1 = -2\xi^3 \qquad \text{for } k = 2$$

(the remaining coordinates are constant; ξ is the parameter along the geodesic).

In order to describe the 'geodesics' for $k = 3$ we have to consider the polynomial

$$\Phi(p_0) = \frac{p_0}{5} + \frac{p_1 p_0^3}{3} + \frac{q_2 p_0^2}{2} + p_2 p_0 + \frac{q_1}{2} = \frac{q_1}{2} + \int F\,dp_0.$$

The 'geodesics' in the space of rays, equipped with the Darboux coordinates $(p_1, \ldots, q_n)$, are defined by the conditions:

$$\Phi \text{ has a root } \xi \text{ of multiplicity } \geq 3,$$

$$p_1; p_3, q_3, \ldots, p_n, q_n \text{ are constant.}$$

The 'upper geodesics' are defined by the same conditions, supplemented by $q_0 = 0$, $p_0 = \xi$.

As an example, let $k = 3$, $n = 2$ (this choice corresponds to a biasymptotic ray on a surface in euclidean 3-space). Then the preceding formulas imply

Theorem 4 ([23]). *A generic Lagrangian subvariety of the cartesian product* M *of a swallowtail and a line in the symplectic manifold* $\mathbf{R}^4$,

$$M = \{p_1, p_2, q_1, q_2 : \Phi' \text{ has a root of multiplicity } \geq 2\},$$

is reducible, by a (formal) symplectic diffeomorphism, to the normal form

$$\{p_1, p_2, q_1, q_2 : \Phi \text{ has a root of multiplicity } \geq 3\}.$$

The last-mentioned variety is the Lagrangian unfurled swallowtail. Hence the preceding theorems imply:

Corollary 5 (*[23]*). *A generic Lagrangian subvariety of the 3-dimensional manifold of 'tangent rays' is diffeomorphic, in a neighborhood of a 'biasymptotic ray', to the unfurled swallowtail.*

Unlike the preceding theorems, this corollary (and its higher-dimensional generalisations) has been proved in the C^∞ and analytic (or holomorphic) situations (see below, §§ 7.3–7.4).

Let us apply corollary 5 to a surface in euclidean 3-space. Consider a pencil (a 1-parameter family) of geodesics on a surface. The straight lines tangent to the geodesics of the pencil form a 2-dimensional singular variety in the 4-dimensional space of straight lines in 3-space.

Corollary 6 (*see [23]*). *At a biasymptotic tangent line, the variety of tangent lines is locally diffeomorphic to a (Lagrangian) unfurled swallowtail (provided that the surface and the pencil are generic).*

The extremal curves of the obstacle problem in 3-space are formed by segments of geodesics of the obstacle and by segments of their tangent lines. The system of shortest paths from a point (or set) to points in the space contains, as a rule, a family of tangent straight lines of the geodesics of a pencil. This family generically contains isolated biasymptotic tangent straight lines. Hence corollaries 5 and 6 describe one generic singularity of a system of extremals in the obstacle problem: the unfurled swallowtail.

The appearance of the unfurled swallowtail as a generic singularity of ray systems in the obstacle problem is in no way a priori evident. It is difficult to understand

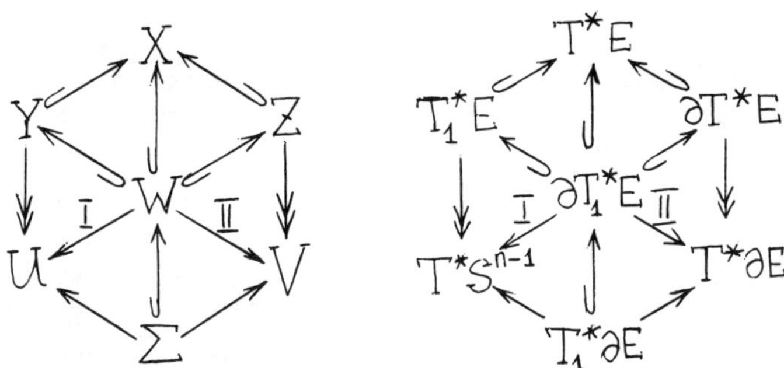

Figure 100: The diagram generated by two hypersurfaces in a symplectic space

why a biasymptotic straight line should be related to a polynomial of degree 5. Even if one knows that such a relation exists, it is difficult to find this polynomial, or to describe it geometrically.

The system of extremals of an ordinary variational problem is a Lagrangian submanifold of the symplectic manifold of extremals. We see that in problems with one-sided constraints the system of extremals is no longer a Lagrangian *manifold*: it acquires singularities, and becomes a Lagrangian *variety*.

The singularities of the Lagrangian varieties thus defined are very special: they are the singularities of Lagrangian subvarieties of the hypersurface of 'tangent rays', which is also singular.

The definition of Lagrangian varieties arising in the obstacle problem is far from the abstract axiomatic constructions which provide extensions of the notion of Lagrangian manifold to the singular case. For instance, one can study Lagrangian ideals (closed with respect to the operation of Poisson bracket), or the varieties defined by generating families violating the transversality condition. These axiomatic constructions lead to another hierarchy of Lagrangian varieties than in the obstacle problem. I think that the latter definition (a Lagrangian variety as a subvariety of the variety of 'tangent rays') is the correct definition of Lagrangian variety, and that the semicubic parabola in the plane and the unfurled swallowtail in symplectic 4-space are the simplest objects of this (future) theory.

We put

X	a symplectic manifold;
Y, Z	two hypersurfaces of X;
W	the intersection $Y \cap Z$;
U, V	the spaces of the characteristics of Y and Z;
Σ	the critical set of the projections $W \to U$, $W \to V$.

The natural inclusions and canonical projections form a hexagonal commutative diagram, as shown in Fig. 100. We use $\hookrightarrow$ for embeddings, $\twoheadrightarrow$ for fibrations. When studying singularities of systems of rays and fronts, it is always useful to keep this diagram in mind.

Example. On the right part of Fig. 100 the general diagram is computed for the particular case corresponding to the usual differential geometry of a hypersurface ∂E in a euclidean space $E = \mathbf{R}^n$. Here,

$$
\begin{aligned}
&X^{2n} &&= T^*E &&= \{x, p\} &&\text{is the symplectic phase space} \\
& && && &&\text{of a free particle;} \\
&Y^{2n-1} &&= T_1^*E &&= \{x, p : p^2 = 1\} &&\text{defines the metric of } E; \\
&Z^{2n-1} &&= \partial T^*E &&= \{x, p : F(x) = 0\} &&\text{defines the hypersurface } \partial E; \\
&U^{2n-2} &&= T^*S^{n-1} &&= &&\text{the manifold of oriented} \\
& && && &&\text{straight lines in } E; \\
&V^{2n-2} &&= T^*\partial E &&= &&\text{the phase space of the} \\
& && && &&\text{constrained particle;} \\
&W^{2n-2} &&= \partial T_1^*E &&= \{x, p : F(x) = 0, p^2 = 1\} = &&\{\text{unit boundary vectors}\}; \\
&\Sigma^{2n-3} &&= \partial T_1^*(\partial E) &&= \{x, p : F(x) = 0, p^2 = 1, p(\partial F/\partial x) = 0\} = &&\{\text{unit boundary tangent vectors}\}.
\end{aligned}
$$

7.2   Contact geometry of pairs of hypersurfaces

The results of § 7.1 have contact variants. In order to formulate them, we use the contactification transforming the symplectic phase space T^*M into the space of 1-jets of functions $J^1(M, \mathbf{R})$. The dimension of the latter space equals that of the phase space plus 1. As a new coordinate we add the value of the generating function. This transforms a Lagrangian submanifold to a Legendre submanifold (at least locally).

Using the Darboux coordinates $(p_0, \ldots, q_n; z)$ in J^1, the contact structure is defined by the form $\alpha = dz - p\,dq$.

We assume that:

1. the intersection of the two hypersurfaces in contact space is transversal;

2. the intersection is transversal to the contact hyperplane;

3. the characteristics of each hypersurface have first order of tangency with the intersection of the hypersurfaces, at all points of interest to us.

Theorem 1 (see [163]). *Locally the pair can be reduced to the Melrose normal form*

$$q_0 = 0, \qquad q_0 = p_0^2 + p_1,$$

using a local C^∞ contactomorphism of the ambient space.

Remark. If either the intersection is not transversal to the contact planes or the order of tangency exceeds 1, then the C^∞ classification involves functional moduli (see [163], [165]).

Melrose's theorem above cannot be applied in a contact 3-space, since the conditions 1–3 are contradictory in this case. The line of intersection of a generic pair of surfaces in a contact 3-space is tangent to the characteristics at isolated points, and generically the order of tangency is 1 but the line is not transversal to the contact planes at these points. Near such points the pair is reducible (by a formal, or C^∞, diffeomorphism) to the normal form (using Darboux coordinates):

$$z = q_0, \qquad p_0^2 = q.$$

The series reducing a pair of analytic surfaces to this normal form are, generically, divergent. Indeed, two involutions are intrinsically defined on the line of intersection, and two such involutions with a common fixed point have functional moduli ([167]).

Theorem 2 (see [164], [165]). *A codimension-1 submanifold of a hypersurface in a contact manifold is locally reducible to one of the forms of theorem 1 in § 7.1, using a formal contactomorphism, provided that the genericity conditions stated below hold.*

Denote by $\Sigma(k)$ the set of points of the submanifold at which the order of tangency with the characteristics is at least k (thus, k denotes the same number as in theorem 1 in § 7.1).

The *nondegeneracy conditions* are:

$k = 1$ the submanifold is transversal to the contact planes;

$k = 2$ $\dim \ker d\alpha|_{\{\alpha=0\} \cap T_x\Sigma(1)} = 1$ $\qquad\qquad\qquad\qquad\qquad$ (*)

$k = 3$ (*) and ($\Sigma(2)$ is transversal to the contact planes).

Generically, these conditions are violated only on a codimension-1 submanifold of $\Sigma(1)$.

Example. Consider a hypersurface in a euclidean space E. Construct a map exp from the manifold of nonzero vectors based at points of the hypersurface into the manifold of unit vectors of the ambient space. This map sends the vector xy, based at the point x, to the unit vector with the same direction, based at the point y.

The map exp may be regarded as the first canonical map for a pair of hypersurfaces in the contact space $J^1(E, \mathbf{R})$. Namely, choose as the first hyperplane the Hamilton—Jacobi equation for particles moving with velocity 1, $(\nabla z)^2 = 1$. The second hypersurface in the contact space is the cylinder over the given hypersurface in euclidean space.

The formulas above give the local normal forms for the map exp.

Theorem 3 (see [164], [165]). *The singularities of the map* exp *at the vectors of asymptotic direction of a generic surface in euclidean 3-space are given by the following table:*

point	$\Pi_{3,1}$	$\Pi_{4,1}$	Π_5	$\Pi_{4,3}$	$\Pi_{3,2}, \Pi_{3,3}$	$\Pi_{4,2}$
singularity	A_2, A_2	A_2, A_3	A_2, A_4	A_3, A_3	A_2	A_3

The types $\Pi_{i,j}$ are defined in § 6.1. The map from A_k is (equivalent to) projection along the x-axis of the surface $\{x, \lambda : x^{k+1} + \lambda_1 x^{k-1} + \cdots + \lambda_k = 0\}$ into the λ-plane.

A unit vector in euclidean space will be called *singular* (with respect to a given generic surface) if the straight line containing this vector is tangent to the surface. The singular unit vectors form a 4-dimensional hypersurface in the 5-dimensional manifold of unit vectors.

Theorem 3 implies

Corollary 4. *The hypersurfaces of singularities of the singular unit vectors at unit vectors of asymptotic tangent lines are: a cylinder over a semicubic parabola (at vectors of ordinary asymptotic tangent lines), a cylinder over a swallowtail (at vectors of biasymptotic tangent lines), and a cylinder over a 3-dimensional swallowtail (at vectors of tri-asymptotic tangent lines.*

Applications to the obstacle problem.

Two types of Legendre varieties are associated to the obstacle problem: varieties of 1-jets of (multivalued) time functions and varieties of contact elements that are tangent to the fronts. The results above yield information about the singularities of these varieties in the plane and in 3-space (the general case is discussed below, in § 7.5).

To obtain this information we first translate to the contact case the constructions of canonical projections (see section C of § 7.1). We will use the terminology of the obstacle problem in the riemannian case, but the whole theory is practically unchanged in the general case of a generic pair of hypersurfaces in an arbitrary contact manifold.

We begin with the manifold $J^1(M, \mathbf{R})$ of 1-jets of functions on a manifold with boundary M, bounded by a smooth hypersurface ∂M. The source point of a jet will be denoted by $x \in M$, the value by $t \in \mathbf{R}$, and the first differential at x by p.

Consider in the contact space of 1-jets on m the two hypersurfaces

$$SJ^1(M, \mathbf{R}), \qquad \partial J^1(M, \mathbf{R}).$$

The first is defined by the Hamilton—Jacobi equation $p^2 = 1$, the second by the condition $x \in \partial M$.

The characteristics of the first hypersurface are the orbits of the geodesic flow in the space of the sphere bundle ST^*M, with the natural parameter t on these orbits. The set of characteristics may be identified with the space of the spherical tangent bundle of M. Indeed, choose a time moment τ. Associate to an arbitrary point of a parametrised orbit of the geodesic flow the point corresponding to the parameter value $t = \tau$. This defines a fibration $SJ^1(M, \mathbf{R}) \twoheadrightarrow ST^*M$. The fibers are the characteristics. Hence we have identified the space of characteristics of $SJ^1(M, \mathbf{R})$ with the manifold ST^*M. This identification depends on the time moment τ.

A characteristic of the second hypersurface is formed by the 1-jets at a point of ∂M of the extensions of a function from ∂M to M. The manifold of characteristics is thus naturally identified with the space $J^1(\partial M, \mathbf{R})$.

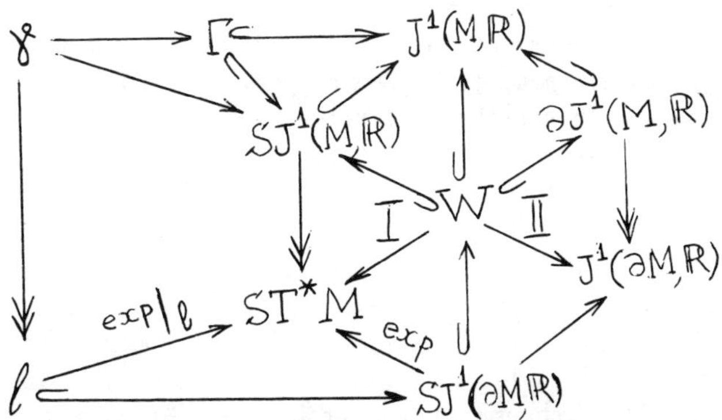

Figure 101: The diagram generated by the obstacle problem

The contact structures of the spaces of characteristics coincide with the standard contact structures in the spaces ST^*M and $J^1(\partial M, \mathbf{R})$; the latter were defined in § 3.1.

The intersection of our 2 hypersurfaces is the light bundle of contact elements over space–time, restricted to the boundary:

$$W = \{(x, p, t) : p^2 = 1, \quad x \in \partial M\}.$$

The first canonical projection $\mathcal{I} : W \to ST^*M$ sends a boundary vector at time t to the unit vector, tangent to the same geodesic, at time τ.

The second canonical projection $\mathcal{II} : W \to J^1(\partial M, \mathbf{R})$ sends the 1-jet of a function on M at a point of ∂M to the 1-jet of the restriction to ∂M of the function. This map has a fold singularity on the hypersurface Σ of W formed by the unit vectors that are tangent to ∂M,

$$\Sigma = \{(x, p, t) : p^2 = p_\partial^2 = 1, \quad x \in \partial M\},$$

p_∂ being the restriction to $T_x \partial M$ of the form p.

The set of critical values of the second canonical projection is the hypersurface $SJ^1(\partial M, \mathbf{R})$ defined on ∂M by the Hamilton—Jacobi equation $p_\partial^2 = 1$. We obtain the embedding $SJ^1(\partial M, \mathbf{R}) \hookrightarrow W$ (the natural diffeomorphism onto Σ) and the map $\exp : SJ^1(\partial M, \mathbf{R}) \to ST^*M$ (in essence, this map coincides with the map $\exp$ defined in the example above theorem 3).

These objects form the hexagonal part of the diagram shown in Fig. 101.

A system of geodesics on ∂M orthogonal to a fixed hypersurface of ∂M and parametrised by the (oriented) distance from this hypersurface, defines a solution of the Hamilton—Jacobi equation $p_\partial^2 = 1$ on ∂M, and hence a Legendre submanifold l that can be embedded in $SJ^1(\partial M, \mathbf{R})$. From this Legendre submanifold we will construct two Legendre varieties: the 1-graph of the time function, and the set of contact elements that are tangent to the fronts in the obstacle problem.

Consider the image of the manifold l under the map $\exp$, and the union Γ of the characteristics of the first hypersurface $SJ^1(M, \mathbf{R})$, which intersects l.

Theorem 5 ([166], [170]). *The varieties* $\exp l$ *and* Γ *are Legendre subvarieties in* ST^*M *and* $J^1(M, \mathbf{R})$, *respectively. Namely, the first one is formed by the contact elements of the front corresponding to the time moment* t, *and the second is formed by the 1-jets of the (multivalued) time function of the obstacle problem, where the obstacle is bounded by* ∂M, *provided that the restriction of the time function to* ∂M *is defined by the Legendre submanifold* l *of* $J^1(\partial M, \mathbf{R})$.

The 1-graph Γ of the time function can be described by using the smooth total space γ of the fibration $\gamma \twoheadrightarrow l$ with 1-dimensional fibers, induced from the fibration $SJ^1(M, \mathbf{R}) \twoheadrightarrow ST^*M$ by the map of bases $\exp|_l : l \to ST^*M$. Namely, Γ is the image of the manifold γ under the induced map of the total spaces, $\gamma \to SJ^1(M, \mathbf{R})$.

Theorems 5 and 2, and theorem 1 and corollary 5 of § 7.1, reduce the whole left part of the diagram shown in Fig. 101 to normal form, provided that the number k in theorem 1 does not exceed 3 (for instance, at the asymptotic and bi-asymptotic rays of the obstacle problem in 3-space). Namely, for the following objects we obtain normal forms:

$$SJ^1(M, \mathbf{R}), \ W, \ SJ^1(\partial M, \mathbf{R}), \ ST^*M, \ l, \ \exp l, \ \mathcal{I}, \ \exp, \ \exp|_l, \ \gamma, \ \Gamma.$$

With the exception of W, $SJ^1(\partial M, \mathbf{R})$, and $\exp$, the objects are reducible to normal form by C^∞ or analytic contactomorphisms for arbitrary values of k (see § 7.5), that is, in the higher-dimensional obstacle problem.

Here I will not write out explicitly all normal forms, and mention only the following:

Corollary 6 ([166], [170]). *In a generic obstacle problem in 3-space, the variety of contact elements of a front and the variety of 1-jets of a multivalued time function have (in a neighborhood of an element corresponding to a bi-asymptotic tangent ray leaving the obstacle surface) singularities diffeomorphic to the singularity at the origin of the variety of polynomials*

$$x^5 + A_1 x^4 + \cdots + A_5 \qquad (x^7 + a_1 x^6 + \cdots + a_7)$$

that have a root of multiplicity at least 3 (respectively, a root of multiplicity at least 5) and for which the sum of the roots is 0:

$$\exp l \simeq \{(x - u)^3 (x^2 + 3ux + v)\},$$

$$\Gamma \simeq \{(x - u)^5 (x^2 + vx + w)\}.$$

On Γ *the time function has the form* $t = \pm a_1 + const;$ *any ray (characteristic) is formed by the shifts of a polynomial along the* x-*axis.*

In the generic planar obstacle problem, in a neighborhood of an element corresponding to an inflectional tangent we have

$$\exp l \simeq \{(x - u)^2 (x + 2u)\} \subset \{x^3 + A_1 x^2 + A_2 x + A_3\},$$

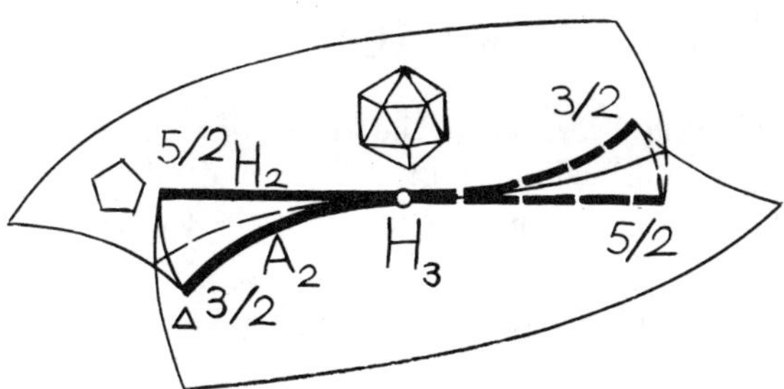

Figure 102: The graph of the multivalued time function regarded as the discriminant of the icosahedron symmetry group H_3

$$\Gamma \simeq \{(x - u)^4(x + v)\}.$$

The expressions of the contact elements of the space of 1-jets and the normal forms of the surfaces $SJ^1(M, \mathbf{R})$ in the coordinates A_i and a_i can be found in § 7.5.

Example. Corollary 6 implies that the variety of contact elements of a front is diffeomorphic to an unfurled swallowtail (in the spatial problem), or to a semicubic parabola (in the planar problem). The 1-graphs of the time functions are diffeomorphic to the cylinders over these varieties.

It is clear from Fig. 5 that in the planar obstacle problem the projection of the 1-graph of the time function to the plane has (at a generic point of inflection of the boundary of the obstacle) a singularity that is diffeomorphic to the projection of the bifurcation diagram of C_3,

$$\{x^3 + ax^2 + bx + c = (x - u)^2(x - v)\} \to \{a, c\};$$

shown in Fig. 92.

The shape of the 0-graph of the time function is also deducible from Fig. 5. Each evolvent should be considered as the planar curve belonging to a horizontal plane at height corresponding to the value at t (this horizontal plane is an isochrone of space–time). The surface in space–time swept by the evolvents has 2 cuspidal edges (of orders 3/2 and 5/2) and a line of selfintersection; these are tangent to each other of order of tangency 3. This surface (Fig. 102) is diffeomorphic to the variety of nonregular orbits of the reflection group H_3 (that is, of the icosahedron symmetry group).

The normal form of this variety is

$$\{(A, B, C) : x^5 + Ax^4 + Bx^2 + C \text{ has a multiple root}\}.$$

The time function on space–time can be reduced to normal form by a diffeomorphism of space–time preserving this surface. The resulting normal form is: $t = A + \text{const}$ (see [1], [4], [5]).

7.3   Unfurled swallowtails

In this section we will study certain special algebraic varieties: the variety of polynomials of a fixed degree and having a root of fixed multiplicity. Such varieties provide normal forms for many singularities appearing in the theory of ray systems.

Example 1. The variety of polynomials

$$x^5 + Ax^3 + Bx^2 + Cx + D$$

having a root of multiplicity (at least) 3 is called the *open* (or *unfurled*) *swallowtail*.

This surface is locally diffeomorphic to the surface in the 4-space of straight lines in euclidean 3-space formed by the extremal rays of a generic obstacle problem in 3-space. Namely, the vertex of the unfurled swallowtail corresponds to the ray tangent to the curve along which the directions of the outgoing rays are asymptotic for the surface of the obstacle (see Fig. 95 and its accompanying text).

Example 2. Consider a wave front that generically propagates in 3-space. The cuspidal edges of the fronts sweep the caustic. Assume that the caustic has a swallowtail (this singularity occurs on generic caustics). The moving front is described by a hypersurface (the big front) in space–time. In our case this hypersurface has a 3-dimensional swallowtail singularity (diffeomorphic to that of the variety of polynomials $x^5 + Ax^3 + Bx^2 + Cx + D$ having a multiple root) The cuspidal edges of the momentary fronts sweep a 2-dimensional variety on the big front. It is formed by those points (A, B, C, D) that correspond to the polynomials having a root of multiplicity (at least) 3.

Thus, *the cuspidal edges of the momentary fronts sweep an unfurled swallowtail in 4-dimensional space–time.*

Differentiation of a polynomial decreases by 1 both its degree and the multiplicity of each root. The unfurled swallowtail in the 4-space of polynomials of degree 5 is projected by this map onto the usual swallowtail in the 3-space of polynomials of degree 4 (Fig. 10). This projection has an interpretation in physics. Space–time is fibered into the world lines of events 'at a given point of space'. The base of this fibration is 3-dimensional. A generic big front in space–time can be reduced to the normal form

$$\{(A, B, C, D) : x^5 + Ax^3 + Bx^2 + Cx + D \text{ has a multiple root}\}$$

by a diffeomorphism transforming the world lines into straight lines parallel to the D-axis. (Moreover, by Ljashko's theorem [159], see also § 6.3, this diffeomorphism may be chosen such that on each world line the time differs from D by an additive constant.)

Thus, *differentiation of polynomials corresponds to projection of space–time along world lines*: the big front lives in the space of polynomials of degree 5, and the caustic lives in the space of polynomials of degree 4 (using normal forms, that is, if we choose convenient local coordinates in a neighborhood of the singularity).

The study of the sweeping of the caustic by the cuspidal edges of momentary fronts is now divided into two separate problems: we first have to study the sweeping of the unfurled swallowtail on the big front, and then we project the result onto the ordinary swallowtail.

To study the sweeping of the unfurled swallowtail on the big front, we have to reduce to normal form the time function. *A generic time function on the whole big front is reducible to the normal form*

$$t = \pm A + const$$

by a diffeomorphism preserving the big front (see [1]). However, this diffeomorphism destroys the world lines: it transforms a family of straight lines parallel to the D-axis into a family of curves. Hence we cannot use this normalising diffeomorphism in the study of sweeping of the caustic. Thus we are led to the problem of reducing the time function to normal form using a diffeomorphism that preserves the fibration of space–time into world lines.

It is impossible to require that the reducing diffeomorphism preserves the whole big front. Indeed, such diffeomorphisms form a very small set, not large enough to reduce to normal form the family of cuspidal edges of momentary fronts. Hence we will reduce to normal form only the *restriction* of the time function to the variety formed by the cuspidal edges of the momentary wave fronts in space–time. This variety has an unfurled swallowtail singularity at the origin.

Let us state the final results.

Definition 1. A *time function* is a function of (A, B, C, D) that vanishes at the origin and has at the origin a nonzero derivative along the A-axis.

Definition 2. Two time functions are *equivalent* if their restrictions to the unfurled swallowtail can be transformed into each other by a diffeomorphism of the (A, B, C, D)-space preserving the unfurled swallowtail and the fibration into straight lines parallel to the D-axis, followed by a diffeomorphism of the time axis.

Theorem 1 (see [72]). *A generic time function is formally equivalent to the function* $A + D$.

In other words, the family of cuspidal edges of the momentary fronts is locally reducible to the normal form $A + D = $ const, at least at the level of formal series.

Remark 1. Such a reduction is certainly impossible at the holomorphic level: generically the reducing series are divergent.

Indeed, the cuspidal edge of a moving wave front passes twice through a point on the line of selfintersection of the caustic (Fig. 103).

[Theorem 1 implies that the time interval between these moments scales like $s^{5/2}$, where s is the distance to the vertex of the swallowtail.]

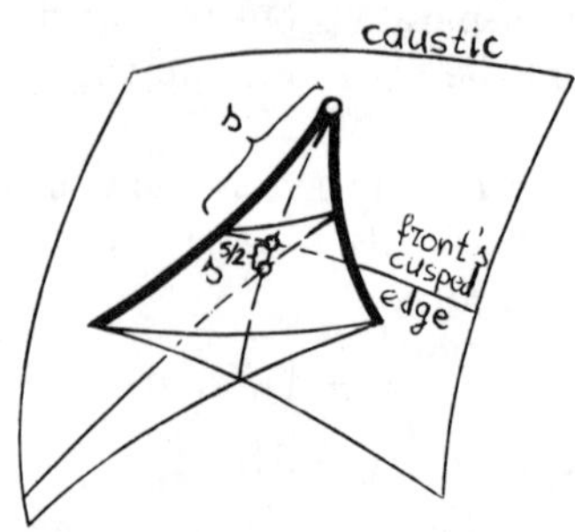

Figure 103: Sweeping the caustic of the swallowtail by the cuspidal edges of moving wave fronts

Consider the pre-image of the line of selfintersection of the swallowtail on the surface of the unfurled swallowtail. It is a smooth curve, containing the vertex of the unfurled swallowtail. On it, two involutions preserving this vertex are naturally defined: one interchanges any two points on the big front that are mapped to the same point of selfintersection of the caustic, the other interchanges any two points belonging to the same isochrone.

A pair of germs of involutions of $\mathbf{C}$ at a common fixed point can be reduced to simple normal form at the level of formal series. In the holomorphic case, however, these series are generically divergent. Moreover, pairs of involutions have functional moduli (Voronin [167], [168], Ecalle [169]).

Thus, the theory of pairs of involutions of a line implies *divergence* of the series reducing the time function on the unfurled swallowtail to the normal form $A + D$.

Remark 2. Other theorems on normal forms of sweeping various surfaces by the cuspidal edges of moving wave fronts are proven in [72], [170], [144].

Example 1. In a neighborhood of the 'lips' or 'beak-to-beak' perestroikas of a caustic (or front) moving in the plane, *the moving curve may be reduced, by a formal diffeomorphism of space–time which is fibered over the time axis, to the normal form* [72]

$$t = u + u^{3/2} \pm v^2$$

(where u and v are the coordinates in the plane and t is the time coordinate). By forbidding transformations of time, we may reduce the family to the normal form

$$t = f(u) + u^{3/2} \pm v^2, \qquad f'(0) = 0,$$

by a *true* (*genuine*, not only formal) diffeomorphism.

Reduction of a holomorphic family of caustics or wave fronts to the first normal form can only be formal: generically the reducing series are divergent.

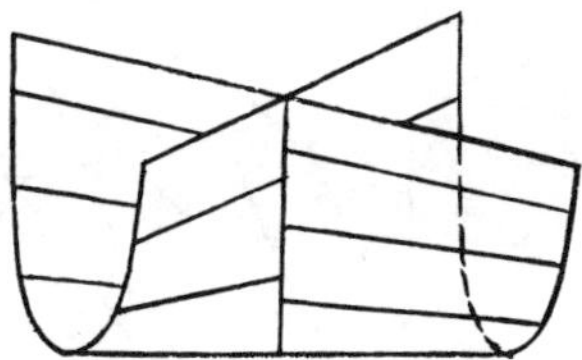

Figure 104: Sweeping the umbrella by smooth curves

Example 2. Consider the sweeping of a Whitney—Cayley umbrella (the image under the map $(p,q) \mapsto (p, pq, q^2)$) by the cuspidal edges of the momentary caustics or wave fronts. *The normal form of the time function is* [72]:

$$t = q + a(q^2).$$

Reduction to this normal form can be done using a genuine (not just a formal) diffeomorphism, provided diffeomorphisms of the time axis are not allowed. If the latter are allowed, the normal form is $t = q$. For instance, *a generic family of smooth curves sweeping the umbrella can be reduced to a family of straight lines by a diffeomorphism preserving the umbrella* (Fig. 104).

Consider a caustic moving in space (*not* in space–time). The cuspidal edges of the moving caustic sweep a surface, called a bicaustic in [72].

Definition. The *bicaustic* (of a caustic, being a hypersurface in the total space of a fibration with 1-dimensional fibers) is the projection to the base space of the fibration of the union of the singularities (other than selfintersections) of the initial caustic.

Let us describe the *classification of the singularities of generic bicaustics in the plane and in 3-space* (details are published in [72]; the results first appeared in [13]).
A generic bicaustic in the plane is, in a neighborhood of each of its points (taken to be the origin), diffeomorphic to one of the following curves:

$$u = 0, \quad uv = 0, \quad u^3 = v^2, \quad u^3 = au^2v^2 + v^6 \qquad (4a^3 + 27 \neq 0).$$

These singularities are stable (the deformation can only change the value of the parameter a in the last normal form).
In 3-space, a generic bicaustic can only have the following singularities (besides selfintersections):

i) cuspidal edges (formed by A_4 points);

ii) a swallowtail (an A_5 point);

iii) an umbrella (some A_3 points);

iv) D_4 lines with D_5 points;

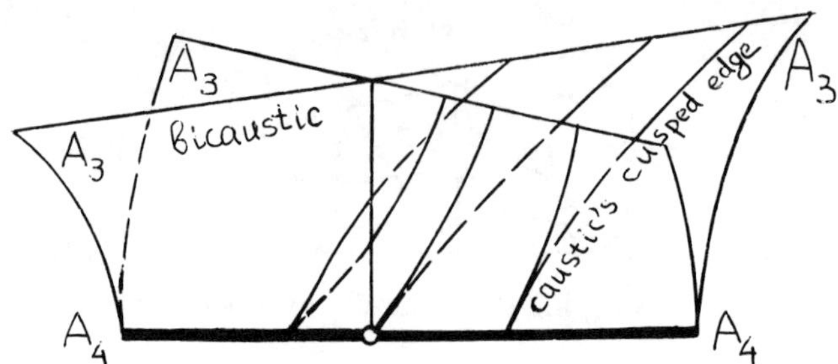

Figure 105: The folded umbrella; the A_4 singularity of the bicaustic

v) intersections of the lines A_4 and D_4 with lines of selfintersection.

Case A_3. *A generic projection of the surface of singularities of a generic caustic in 4-space to 3-space is, in a neighborhood of any regular (A_3) point of this surface, either a local embedding or equivalent to the standard map from the plane onto the umbrella,*

$$(u, v) \mapsto (u, uv, v^2)$$

(in the normal form above, this point corresponds to the origin).

The stability of both maps is well known.

Thus, the only singularities of a generic bicaustic in 3-space are (next to transversal selfintersections) Whitney—Cayley umbrellas and singularities on the lines swept by the point singularities of momentary caustics (that is, A_4 and D_4 lines).

Case A_4. *A generic A_4 bicaustic in 3-space is locally diffeomorphic to one of the following 3 surfaces:*

$$u^2 = v^3, \quad u^2 = v^3 w^2, \quad u^2 w = v^3 w$$

(the A_4 point corresponds to the origin, A_5 and D_5 points do not occur). These singularities are stable.

Thus, the projection of the line A_4 is a semicubic cuspidal edge of the bicaustic, except at the (isolated) points of the cuspidal edge at which both parts of the bicaustic separated by the cuspidal edge intersect each other. The singularity of the bicaustic at these points is the folded umbrella (Fig. 105).

The proof is base on a general theorem: *the image of a generic surface having a semicubic cuspidal edge under the folding map $((u, v, w) \mapsto (u^2, v, w))$ of 3-space is locally diffeomorphic to the 'folded umbrella' $u^2 = v^3 w^2$.*

Case D_4. *A generic D_4-bicaustic in 3-space is diffeomorphic to either the plane (the case D_4^+, the purse) or the surface $w(w - v^2)(w - Av^2) = 0$, where $A = a + u$, $a \neq 0$, at generic points of the D_4 curve, and $A = u$ or $A = a \pm u^2$, $a \neq 0, 1$, at certain special points (the case D_4^-, the pyramid, Fig. 106).*

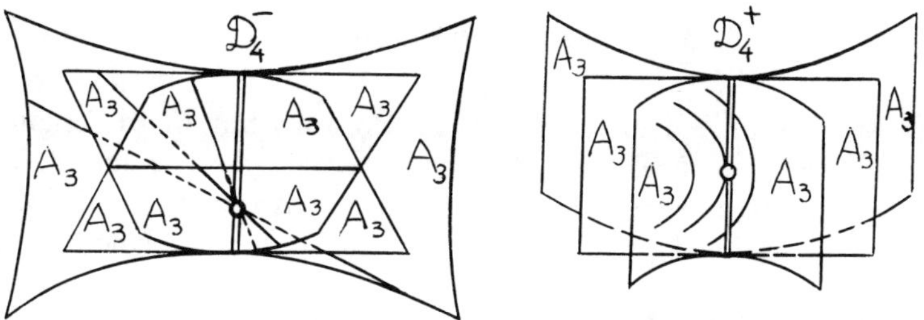

Figure 106: The D_4 singularity of the bicaustic

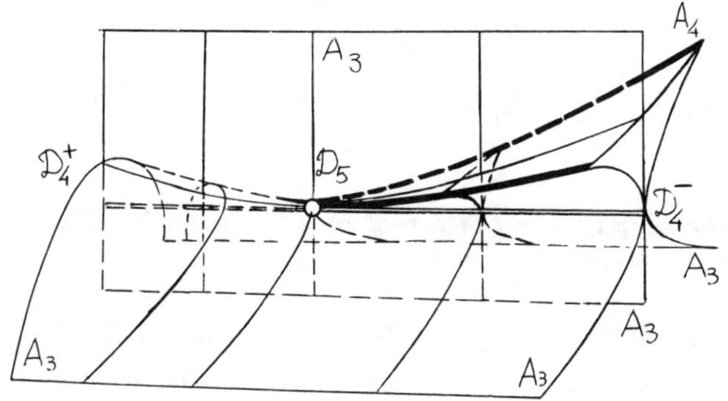

Figure 107: The D_5 singularity of the bicaustic

These singularities are also stable (the value of the parameter a may change under deformation).

Case A_5. *A generic bicaustic in 3-space is diffeomorphic to the swallowtail at the A_5 points.*

Case D_5. The normal form of the bicaustic in 3-space at its D_5 points is unknown (Fig. 107).

Of course, a generic bicaustic also contains points of transversal intersection of different branches of singularities, as described above.

We return to the unfurled swallowtail. Theorem 1 describes the generic sweeping of the bicaustic by the cuspidal edges of the momentary fronts at the A_5 singularity.

The proof of theorem 1 depends on the study of vector fields tangent to the unfurled swallowtail.

Theorem 2. *A vector field in 3-space that is tangent to the swallowtail and holomorphic at the vertex is the projection of a vector field in 4-space that is tangent to the* *unfurled swallowtail and holomorphic at the vertex.*

This theorem is proved in [72] by straightforward but weird computation: the result depends on the cancellation of a lot of terms by 'sheer luck'. The analysis of the reasons for these cancellations has led A.B. Givental to the following general theory of multidimensional unfurled swallowtails in spaces of polynomials of arbitrary degree.

Consider *Givental's tower*, formed by the spaces of polynomials of arbitrary degree, with complex coefficients, and with leading coefficient 1 (a similar theory exists for the real polynomials).

The nth floor of the tower is the space

$$\mathbf{C}^n = \{x^n + a_1 x^{n-1} + \cdots + a_n\}.$$

Differentiation followed by division by the degree projects each floor onto the preceding one (the projection is a fibration with 1-dimensional fibers).

Definition. The *comultiplicity* of a root of a polynomial is the difference between the degree and the multiplicity.

Differentiation does not change the comultiplicity of a root. The polynomials with at least one root of comultiplicity at most k form an (affine) algebraic subvariety of the nth floor. For a fixed k, we obtain a chain of algebraic subvarieties of the spaces of polynomials of varying degree and with at least one root of comultiplicity at most k. Projection of a floor of our tower onto the preceding floor sends the upper variety (of the chain in the floor) onto the lower variety of the chain.

Theorem 3 (see [11]). *The chain above stabilises at the $(2k+1)$st floor (this is the first floor on which the root of comultiplicity at most k is always unique).*

More precisely, *each variety in a higher floor* (that is the variety of polynomials of degree $n > 2k + 1$ having a root of comultiplicity at most k) *belongs to a smooth algebraic hypersurface which projects diffeomorphically onto the space of the preceding floor.*

Hence, *all varieties of polynomials of degree n and having a root of comultiplicity at most k are diffeomorphic, provided that $n \geq 2k + 1$* (they are diffeomorphic as singular and as algebraic varieties).

Remark. A similar theorem holds for the tower of polynomials with zero traces (the trace of a polynomial is the sum of its roots). This is an easy consequence of theorem 3, since differentiation commutes with translations.

Example. Consider the projection of the swallowtail, formed by the polynomials

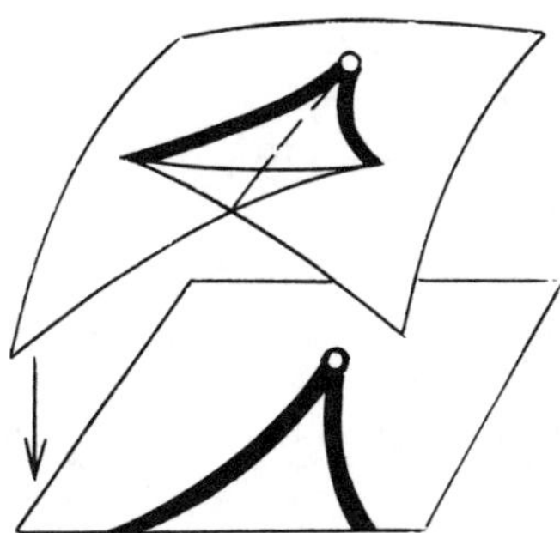

Figure 108: Stabilisation of the 1-dimensional unfurled swallowtail

$x^4 + Ax^2 + B + C$ having a multiple root, onto the (A, B)-plane along the C-axis (Fig. 108). The cuspidal edge of the swallowtail projects diffeomorphically onto the semicubic parabola. Indeed, this cuspidal edge belongs to the surface $C = -A^2/4$. Here $k = 1$ and the chain stabilises at the floor of cubic polynomials.

The smooth hypersurface mentioned in theorem 3 can be explicitly described: it is the graph of a quadratic polynomial. For the first floor above the stable floor $(n = 2k + 2)$ this hypersurface was found by Hilbert [10]. Namely, he proved the easy

Theorem 4. *If*

$$x^n + b_1 x^{n-1} + \cdots + b_n = (x - a)^{k+2}(x^k + a_1 x^{k-1} + \cdots + a_k),$$

then

$$2n! b_n = \sum (-1)^{i+1} i! (n - i)! b_i b_{n-i}, \qquad 1 \le i < n.$$

The equations of paraboloids in spaces of polynomials of higher degree can also be written out explicitly. It is convenient to introduce the differential operators $G^{(s)}$ by the formulas

$$G^{(s)} F = \frac{\frac{d^s F}{dx^s}}{(s - r)!},$$

where r is a fixed number and 'factorials' of negative numbers are defined as the products of the preceding negative numbers, truncated at a suitable far place; for instance, we may use the 'factorials'

$$(s - r)! = (s - r) \cdots (1 - r) \qquad \text{for } s < r.$$

Theorem 5. *If*

$$F = (x - a)^{r+1}(x^{m-1} + \cdots), \qquad \text{where } r > m,$$

then

$$\sum (j - i)G^{(i)}[F]G^{(j)}[F] = 0$$

(summation over $i + j = r + m$, $0 \le i \le m$).

Theorem 5 gives the explicit equation of the paraboloid in the floor of polynomials of degree $r + m$ and containing the variety of polynomials with a root of comultiplicity at most $k = m - 1$.

Theorems 4 and 5 imply theorem 3, and also the following

Corollary. *All n coefficients of the polynomial*

$$(x - a)^r (x^k + a_1 x^{k-1} + \cdots + a_k) = x^n + b_1 x^{n-1} + \cdots + b_n$$

can be expressed as polynomials of the variables $(b_1, \ldots, b_{2k+1})$ with rational coefficients.

Theorems 4 and 5 can be proved by straightforward computation (they describe rather sophisticated combinatorial identities between binomial coefficients). Givental has noted that theorems 3 and 4 can be deduced from the following two identities (in which $F^{(i)} = d^i F/dx^i$):

Theorem 6 (see [11], [171]). *For an arbitrary polynomial F of even degree n the polynomial*

$$R_0[F] = \sum_{i=0}^{n} (-1)^i F^{(i)} F^{(n-i)}$$

is a constant. This constant is 0 if F has a root of comultiplicity smaller than $n/2$.

For an arbitrary polynomial F of odd degree n, the polynomial

$$R_1[F] = \sum_{i=0}^{(n-1)/2} (n - 2i) F^{(i)} F^{(n-i)}$$

is linear, and

$$\frac{d}{dx} R_1[F] = R_0 \left[\frac{dF}{dx} \right].$$

The linear polynomial $R_1[F]$ is identically equal to 0 if F has a root of comultiplicity less than $(n - 1)/2$.

Proof. (See [11].) Consider the 'generalised Newton equation of order $2m$ for a free particle',

$$\frac{d^{2m} u}{dt^{2m}} = 0.$$

The function

$$H = \frac{1}{2} \sum_{i=1}^{2m-1} (-1)^{i-1} u^{(i)} u^{(2m-i)} \qquad \left(= \int u^{(2m)} \dot{u}\, dt \right)$$

is a first integral. Introduce a symplectic structure in the phase space, using the Darboux coordinates

$$q_i = u^{(i-1)}, \quad p_i = (-1)^{i+1} u^{(2m-i)}, \qquad i = 1,\ldots,m$$

(in other words, consider the symplectic structure corresponding to the variational principle $\delta \int L\, dt = 0$, where $L = (u^{(m)})^2/2$).

The Newton equation takes the form of the Hamiltonian system with Hamiltonian H. Its solutions are the polynomials of degree less than $2m$. Theorem 4 and the first part of theorem 6 express the conservation law of H for the equation of a free particle.

Consider the homogeneous and the quasihomogeneous euler vector fields,

$$E_0 = \sum u^{(i)} \frac{\partial}{\partial u^{(i)}}, \qquad E_1 = \sum i u^{(i)} \frac{\partial}{\partial u^{(i)}}.$$

The corresponding dilations transform the phase trajectories of the Hamilton equations of a free particle into phase trajectories. The field

$$V_1 = \frac{2m-1}{2} E_0 - E_1$$

is Hamiltonian (since the homogeneous degree of the symplectic structure is 2 and its quasihomogeneous degree is $2m - 1$).

The Hamilton function of the field V_1 is

$$H_1 = \frac{1}{2} \sum (-1)^{i-1} (2m + 1 - 2i) u^{(i-1)} u^{(2m-i)} \qquad (1 \le i \le m).$$

The Poisson bracket of H and H_1 is readily computed from the homogeneity and quasihomogeneity of H. We obtain

$$\{H, H_1\} = H.$$

This proves the second part of theorem 6 (as well as theorem 3).

Example. Consider the varieties of polynomials $x^n + a_0 x^{n-2} + \cdots + a_{n-2}$ having a root of comultiplicity at most $k = 2$. Such a variety in the space of polynomials of degree 4 is the ordinary swallowtail. In the space of polynomials of degree 5 such a variety is the unfurled swallowtail.

The projection of the unfurled swallowtail onto the ordinary swallowtail is not a diffeomorphism (since the points on the line of selfintersection each have 2 preimages). According to theorem 3, the unfurled swallowtail in the space of polynomials of degree $2k + 1 = 5$ is stable: all varieties in our chain and living in higher floors are diffeomorphic to this variety.

Like in this example, stabilisation always occurs exactly at the lowest floor on which all selfintersections disappear. Hence we will call the first stable variety the *multidimensional open* (or *unfurled*) *swallowtail.* The 1-dimensional unfurled swallowtail is the semicubic parabola. The 2-dimensional unfurled swallowtail is a surface in the space of polynomials of degree 5, etc.

Theorem 7 (see [11]). *Each germ of a holomorphic vector field that is tangent to a multidimensional swallowtail (to the variety of polynomials with multiple roots) can be lifted to a germ of a holomorphic vector field that is tangent to the stable variety over this swallowtail.*

In other words, in the space of polynomials of arbitrary degree $N \geq (2n - 3, n)$ there is a holomorphic germ of any vector field that is tangent to the variety of polynomials with a root of comultiplicity at most $n - 2$ and that projects to a vector field in the space of polynomials of degree n by $N - n$ differentiations.

Remark 1. Similar theorems hold for spaces of polynomials with no restrictions on the leading coefficient and trace, and also for polynomial vector fields in the whole space.

Remark 2. The fields that are tangent to the (multidimensional) swallowtail are well known (see § 4.1). The proof of theorem 7, published in [11], provides explicit formulas for the lifted fields.

Remark 3. In theorem 7 the stability restriction $N \geq 2n-3$ is essential: the assertion of the theorem is wrong for $N = 6$, $n = 5$.

Remark 4. Any polynomial vector field in the space of polynomials of degree n can be decomposed as the sum of a polynomial vector field lifted from the space of polynomials of degree $n - 1$ and a field that is tangent to the variety of polynomials with a multiple root. Hence the following converse to theorem 7 holds: any polynomial vector field that is tangent to the variety of polynomials of degree n and with a root of comultiplicity at most k can be decomposed into the sum of liftings of fields tangent to the preceding swallowtails (i.e. to the varieties of polynomials of degrees $n, n - 1, \ldots, k - 2$ and with a multiple root).

In the sequel we will need one more result on the unfurled swallowtail.

Theorem 8 (see [11]). *The k-dimensional unfurled swallowtail*

$$\{x^n + a_0 x^{n-2} + \cdots + a_{n-2} = (x - t)^{k+1}(x^k + \cdots)\}$$

is a (singular) Lagrangian variety in the space of polynomials of degree $n = 2k + 1$, with leading coefficient 1 and trace 0, equipped with the natural symplectic structure *(see § 1.1, example 10).*

Proof. Our space is the space of characteristics of the hypersurface $h = 0$ in the symplectic space of polynomials of degree $n + 1$, where h is the Hamiltonian of the shifts along the x-axis.

The unfurled swallowtail of dimension k is the projection of the manifold of polynomials of degree $n + 1$ that are divisible by x^{k+2} to the space of characteristics. The symplectic form of the $(2k + 2)$-dimensional symplectic space vanishes on this smooth k-manifold (since this manifold is the coordinate q-plane in the Darboux coordinates of § 1.1). Hence its projection to the $2k$-dimensional space of characteristics is a Lagrangian subvariety (in general, singular).

The explicit expression of the symplectic structure of theorem 8 is easily obtainable from the algorithm of § 1.1.

Example. Consider the ordinary unfurled swallowtail $(k = 2)$

$$\{x^5 + Ax^3 + Bx^2 + Cx + D = (x - t)^3(x^2 + 3tx + s)\}.$$

We start from the usual symplectic structure in the space of polynomials of degree $2k + 3 = 7$,

$$\{q_0 e_7 + q_1 e_6 + q_2 e_5 + q_3 e_4 - p_3 e_3 + p_2 e_2 - p_1 e_1 + p_0 e_0\},$$

where $e_j = x^j / j!$ and (p, q) are the Darboux coordinates of § 1.1. The hyperplane $q_0 = 1$ is fibered into characteristics parallel to the p_0-axis. The space of characteristics is naturally identified with the space of derivatives, that is, with the space of polynomials of degree $2k + 2 = 6$,

$$\{e_6 + q_1 e_5 + q_2 e_4 + q_3 e_3 - p_3 e_2 + p_2 e_1 - p_1\}$$

(equipped with the symplectic structure $\sum dp_i \wedge dq_i$). The Hamiltonian of the shifts is the function

$$H = p_1 + p_2 q_1 + p_3 q_2 + \frac{q_3^2}{2}.$$

Choose the coordinates $(q_1, q_2, q_3, p_3, p_2)$ on the hypersurface $H = 0$. A characteristic of this hypersurface is a polynomial with $H = 0$, together with its shifts along the x-axis. The space of characteristics can thus be identified with the space of polynomials of degree $2k + 2 = 6$ having $H = 0$ and with vanishing trace ($q_1 = 0$). Choose the coordinates (p_2, q_2, p_3, q_3) on this space.

The above space of polynomials can be diffeomorphically projected onto a space of polynomials of degree $2k + 1 = 5$ by differentiation. Hence we have identified the space of characteristics with the space of polynomials of degree $2k + 1 = 5$, with fixed leading coefficient and zero trace. The symplectic structure of the space of characteristics can be written in terms of the other (i.e. nonleading) coefficients of these polynomials. In terms of the chosen Darboux coordinates the expressions for the coefficients are:

$$A = 20q_2, \quad B = 60q_3, \quad C = -120p_3, \quad D = 120p_2.$$

Hence theorem 8 says that the *unfurled swallowtail is a Lagrangian variety for the symplectic structure* $3\,dA \wedge dD - dB \wedge dC$.

In the general case the symplectic structure is the form

$$\sum (-1)^i i!\,j!\,da_i \wedge da_j, \qquad i + j = n - 2.$$

The unfurled swallowtail describes also the generic singularities in the theory of duality of curves in projective space.

Let $f : \mathbf{R} \to \mathbf{R}^3$ be a smooth map. We will call it a *curve of order* (a, b, c) *at the origin* if the first derivative of f that does not vanish at the origin has order a, the first subsequent derivative that is not collinear at the origin with the ath derivative at the origin has order b, and the first subsequent derivative that is not coplanar at the origin with the derivatives of orders a and b at the origin has order c.

Curves of order $(a_1, \ldots, a_n)$ in the spaces $\mathbf{R}^n$, $\mathbf{R}P^n$, $\mathbf{C}^n$, $\mathbf{C}P^n$ are defined similarly. A curve of finite order has an osculating hyperplane at each point.

Definition. The curve $f^* : \mathbf{R} \to \mathbf{R}P^{n*}$ *dual to* $f : \mathbf{R} \to \mathbf{R}P^n$ is formed by the osculating hyperplanes of the curve f.

The following theorem is easy, and well known (see, for instance, [148], and [143], [14], [172]–[174]):

Theorem 9. $f^{**} = f$; *the order dual to* (a, b, c) *is* $(c - b, c - a, c)$. *In* n *dimensions the dual order is given by:*

$$(a_1, \ldots, a_n)^* = (a_n - a_{n-1}, \ldots, a_n - a_1, a_n).$$

The regular points of the curve are distinguished by the condition $a_1 = 1$. The initial part of the *hierarchy of smooth curves* in $\mathbf{R}^3$ is:

$$(1,2,3) \longleftarrow (1,2,4) \longleftarrow (1,2,5) \longleftarrow \cdots$$

$$\text{ordinary point} \qquad \text{flattening} \qquad \text{biflattening}$$

$$(1,3,4) \longleftarrow \cdots$$

$$\text{inflection}$$

The codimension of the order is equal to

$$c = \sum (a_i - i)$$

(the order of codimension c is unavoidable in the family of curves depending on $c - 1$ parameters).

Definition. The variety of tangent hyperplanes (i.e. hyperplanes containing a tangent line) of a curve f in the dual projective space is called the *front* of the Legendre singularity of f (for short, the *front of the curve f*). The front is a developable surface whose cuspidal edge is the dual curve f^*.

Example. Suppose f has the simplest singularity of codimension 1 (that of order $(1, 2, 4)$). Then f^* has a singularity of order $(2, 3, 4)$, similar to the singularity of the curve $(x = t^2,\ y = t^3,\ z = t^4)$. The tangent straight lines of this curve sweep a swallowtail surface—the front of the curve f.

A generic curve in 3-space may have isolated flattening points, but it cannot have a point of another order. Hence the front of a generic curve is a surface whose only singularities (next to selfintersections) are semicubical cuspidal edges and isolated swallowtail singularities.

Consider a generic 1-parameter family of curves in 3-space. The curves corresponding to certain special parameter values may have biflattening points. Consider the family of dual curves. Put each dual curve in its own 3-space, depending on the parameter of the family. These dual curves then sweep a surface in the 4-space $\mathbf{R}^3 \times \mathbf{R}$ (where $\mathbf{R}$ denotes the parameter axis).

Theorem 10 (see [23], [143]). *The surface sweeped by the dual curves is locally diffeomorphic to the unfurled swallowtail. The decomposition of this surface into dual curves is diffeomorphic to the decomposition into the lines $A = const$ of the variety of polynomials $x^5 + Ax^3 + Bx^2 + Cx + D$ having a root of multiplicity at least 3.*

For certain isolated parameter values the curves in the initial family may have an isolated point of inflection. According to O.P. Shcherbak ([143], [172]), the normal form of the surface sweeped by their dual curves is

$$\{x^4 + Ax^3 + Bx^2 + Cx + D \text{ has a multiple root}\}.$$

The decomposition into dual curves is $B = const$.

Among other things, Shcherbak's paper [143] contains the following results:

1. *The front of the curve dual to a generic curve* (that is, the developable surface whose cuspidal edge is the initial smooth curve in $\mathbf{R}^3$ or $\mathbf{R}P^3$) *has a folded umbrella singularity at the points of inflection $(1, 2, 4)$ of the initial curve*. Hence it is locally diffeomorphic to the surface $y^2 = x^2 z^3$ (Fig. 105).

2. *The only singularities of the front of a generic curve in $\mathbf{R}P^n$ are multidimensional swallowtails* (locally diffeomorphic to the discriminants of $A_k,\ k \leq n$). *The union of the front of the curve with the hyperplane dual to the corresponding flattening point of the initial curve is locally diffeomorphic to the discriminant of B_k.*

3. *Consider generic 1-parameter families of curves in $\mathbf{R}P^n$. For certain isolated*

parameter values the singularities of orders

$$(1, \ldots, n-1, n+2) \quad and \quad (1, \ldots, n-2, n, n+1)$$

occur. The normal forms of the corresponding perestroikas of the fronts are:

$$\{x^{n+2} + u_1 x^n + \cdots + u_{n+1} \text{ has a multiple root}\},$$

fronts: $u_1 = const$;

$$\{x^{n+1} + u_1 x^n + \cdots + u_{n+1} \text{ has a multiple root}\},$$

fronts: $u_2 = const$.

The theory of fronts of curves can be viewed as part of the theory of Legendre singularities.

More generally, fronts of submanifolds of arbitrary dimension in a projective space are defined by the following construction.

A submanifold of a projective space defines a Legendre submanifold of the space of contact elements of the projective space: it is formed by those contact elements that contain a tangent space of the initial submanifold. The space of contact elements of a projective space is fibered over the dual projective space (associate with a contact element the hyperplane containing it). This fibration is a Legendre fibration (see § 3.1, Fig. 35). The Legendre submanifold formed by the tangent contact elements of the initial submanifold defines a Legendre map to the dual projective space. The image of this map (i.e. the set of hyperplanes tangent to the initial submanifold) is the front of this Legendre map. We will call it, for short, the *front of the initial submanifold.* The Legendre map is called the *frontal map* (associated with the submanifold).

The front of a submanifold of arbitrary dimension in a projective space is a hypersurface in the dual space.

It is well known that the (higher-dimensional) swallowtails are developable surfaces. The generalisation of this fact to other fronts is the following theorem.

Theorem 11 (see [172]). *Any stable germ of a Legendre map of corank m is Legendre equivalent to a germ of a frontal map associated with an m-dimension submanifold of a projective space. The frontal maps associated with generic curves have only singularities A_k. The perestroikas of Legendre maps of corank m occurring in generic families with finite-dimensional parameter spaces are Legendre equivalent to the perestroikas of the frontal maps associated with the m-dimensional submanifolds of projective spaces.*

7.4   Symplectic triads

The obstacle problem has led to the unfurled swallowtail singularity: a singularity of the Lagrangian variety of rays leaving the surface of the obstacle, in the symplectic space of all rays. It also describes the singularities of two types of Legendre varieties:

the first is formed by the contact elements of a front, the second by the 1-jets of the multivalued time function (§ 7.2).

In § 7.3 we have studied the higher-dimensional unfurled swallowtails, using the geometry of spaces of polynomials and binary forms. We will now use the results of the preceding 2 sections in order to explain why the varieties of polynomials with a root of high multiplicity describe the singularities of Lagrangian and Legendre varieties in higher-dimensional variational problems with one-sided constraints.

The theory of symplectic triads, which is crucial in the sequel, is due to A.B. Givental [8], who obtained their definition by discarding superfluous parts in Melrose's construction described in § 7.1.

Definition 1. A *symplectic triad* (H, L, l) consists of a hypersurface H in a symplectic manifold and a Lagrangian submanifold L that is tangent to H of order 1 along a submanifold l of codimension 1 in L.

Example 1. Consider a domain D on a hypersurface in a riemannian manifold M.

A *bundle of geodesics* on D is defined by the *time function* $s : D \to \mathbf{R}$ (which is equal to the distance from some initial front in D). The time function satisfies the Hamilton—Jacobi equation $(\nabla s)^2 = 1$ (if s is smooth in D).

With this situation we associate a symplectic triad in the phase space $T^* M$. The hypersurface H is formed by all unit vectors ($p^2 = 1$). The submanifold L consists of all extensions of the algebraic 1-forms $p = ds|_q$ from the tangent spaces of the hypersurface D to the tangent spaces of the ambient manifold M.

The hypersurface H intersects the manifold L along a submanifold l of codimension 1 in L: l consists of the extensions that vanish on the normals to D.

[If we identify 1-forms on M with tangent vectors, using the riemannian metric, we can describe l as the graph of grad s, and L consists of the vectors whose orthogonal projections to the tangent spaces of D belong to l.]

Theorem 1. *The triple* (H, L, l) *constructed from a bundle of geodesics is a symplectic triad.*

Proof. The order of tangency of H and L along l is 1, since the riemannian metric is quadratically convex (this order may be higher in less regular variational problems, or in the complex domain).

L is a Lagrangian submanifold, since

$$i^* p \, dq = d\pi^* s$$

(where $i : L \hookrightarrow T^* M$ is the embedding, $\pi : L \to D$ the natural projection, $p \, dq$ the canonical 1-form on $T^* M$).

Hence (H, L, l) is a symplectic triad.

Now we will consider an arbitrary symplectic triad.

Definition 2. The *Lagrangian variety generated by the triad* (H, L, l) is the image of the submanifold l in the space of characteristics of the hypersurface H.

It is clear that this variety is a Lagrangian submanifold at the points at which it is nonsingular.

Theorem 2. *The Lagrangian variety generated by the triad of theorem 1 consists of the rays (the geodesics of the ambient manifold) that are tangent to the geodesics of the initial bundle of geodesics on the hypersurface.*

Indeed, the space of characteristics of the hypersurface $p^2 = 1$ is the space of (oriented) geodesics of the ambient manifold. The initial points, belonging to l, are unit vectors that are tangent to the lines of the initial bundle of geodesics on the hypersurface.

The Lagrangian varieties generated by equivalent symplectic triads are symplecto-morphic. Theorems 1 and 2 imply that in order to find a normal form of the variety of rays leaving the geodesic lines of the bundle, it suffices to reduce to normal form the triad, using a symplectomorphism. The normal forms are provided by the following

Example 2. Consider the symplectic space V of nonhomogeneous polynomials in x, of degree $d = 2k + 2$ and with fixed leading coefficient, equipped with its natural symplectic structure (see § 1.1). The polynomial corresponding to the Darboux coordinates (p, q) is

$$e_d + q_1 e_{d-1} + \cdots + q_{k+1} e_{k+1} - p_{k+1} e_k + \cdots + (-1)^{k+1} p_1 e_0,$$

where $e_j = x^j / j!$.

We denote by h the Hamilton function of the shifts,

$$h = p_1 + q_1 p_2 + \cdots + q_k p_{k+1} + \frac{q_{k+1}^2}{2}.$$

Let H be the hypersurface $h = 0$. Let L be the Lagrangian plane $p = 0$. The hypersurface H is (first order) quadratically tangent to the Lagrangian plane L along the hypersurface $l \subset L$ defined by the equations

$$l: \quad p = 0, \; q_{k+1} = 0.$$

One readily checks the following

Theorem 3. *The Lagrangian variety generated by the triad in example 2 is symplec-tomorphic to the k-dimensional unfurled swallowtail, that is, to the variety of polynomials having a root of multiplicity exceeding k in the symplectic space of polynomials of degree $2k + 1$ (with fixed leading coefficient and of trace 0).*

The main result on triads is the following

Theorem 4. *The germ at the origin of the triad in example 2 is stable. The germ of a generic triad at a point of quadratic tangency of the hypersurface and the Lagrangian submanifold is symplectomorphic to the triad in example 2 or to its suspension.*

The *suspension* over the triad (H, L, l) in $\mathbf{R}^{2a}$ with Darboux coordinates (p, q) is the triad in $\mathbf{R}^{2a+2a'}$ with Darboux coordinates $(p, p'; q, q')$ for which $H_{new} = H + \mathbf{R}^{2a'}$, $L_{new} = L + L'$, where L' is the Lagrangian plane $p' = 0$ in $\mathbf{R}^{2a'}$.

Corollary 5. *The Lagrangian variety generated by a generic triad is locally diffeomorphic, at each point, to the cartesian product of a (higher-dimensional) unfurled swallowtail and a euclidean space.*

Corollary 6. *The germ of the variety of rays leaving the surface of an obstacle is locally diffeomorphic to the cartesian product of a (higher-dimensional) unfurled swallowtail and a euclidean space (provided that the surface of the obstacle and the bundle of geodesics on it are generic).*

Remark 1. The genericity condition in theorem 4 can be formulated in terms of the field of characteristic directions of H. At the points of l the characteristic direction is tangent to L. Thus we obtain on l a field of directions (tangent to L). The genericity condition on the triad in theorem 4 requires genericity of this field: it should be reducible to the local normal form $\partial/\partial x$ by a diffeomorphism of L sending l to the hypersurface
$$x^a + u_1 x^{a-2} + \cdots + u_{a-1} = 0.$$
An equivalent normal form is: l: $q_a = 0$, field: $\partial/\partial q_1 + q_1 \partial q_2 + \cdots + q_{a-1} \partial/\partial q_a$.

Remark 2. Theorem 4 and its corollaries hold in the smooth and in the holomorphic (analytic) cases.

Proof of theorem 4. (See [170], [8].)

1. Rectify H by choosing Darboux coordinates such that the equation of H in a neighborhood of the origin becomes $q_1 = 0$.

2. Choose a Lagrangian space which is linear in these coordinates and which is transversal at the origin to H and L. The parallel spaces form a Lagrangian fibration (in a neighborhood of the origin). The Lagrangian manifold L is a section of this fibration.

3. Choose new Darboux coordinates (linear with respect to the previous ones) such that the fibers have the form $q = $ const. The equation of L now takes the form $p = \partial s/\partial q$; H is still a hyperplane.

4. Introduce the shifted Darboux coordinates $P = p - \partial s/\partial q$, $Q = q$. Now L becomes the zero section of the fibration $(P, Q) \mapsto Q$. We identify this Lagrangian

fibration (locally) with the cotangent bundle T^*L (the point (P, Q) corresponds to the 1-form $P\,dQ$ on th tangent space of L at Q).

5. The intersection of H with a fiber of T^*L is an affine hyperplane in this fiber.

6. An affine hyperplane in the dual space of a linear space is determined by a straight line, containing the origin, in the initial space and a linear 1-form on this line. Indeed, the hyperplane $Y = \{y : y|_x = c\}$ determines the line $\{x : y|_x = z|_x$ for all $y, z \in Y\}$ and the 1-form $y|_x$ on this line. The pair (line, 1-form) determines the hyperplane, and any such pair determines an affine hyperplane.

7. Thus, the hypersurface H can be described in the interior geometry of L by the pair formed by a direction field on L and a field of 1-forms on these directions. In other words, H can be described by the pair formed by a direction field on L and a family of differential 1-forms along the integral curves of the field.

   The hypersurface l is formed by those points of L at which the forms in this family vanish. At the points of l, the directions of the field are characteristic for H.

8. The embedding of the hypersurface l into L, and the fibration of L into integral curves define a projection of the hypersurface along the 1-dimensional fibers. The singularities of this projection are Whitney singularities, provided that the triad is generic (see § 7.2 and remark 1 above).

9. Hence we can choose coordinates (x, u) in L such that the integral curves become $u = \mathrm{const}$, while the hypersurface l will be defined by the equation

$$F = 0, \qquad \text{where } F = x^a + u_1 x^{a-2} + \cdots + u_{a-1}.$$

10. The family of differential 1-forms along the integral curves, written in these coordinates, takes the form $f(x, u)\,dx$. The function f vanishes on l. The order of vanishing is 2, since H is quadratically tangent to L along l. Thus,

$$f(x, u) = g(x, u)F^2(x, u), \qquad \text{where } g(0,0) \neq 0.$$

11. **Lemma.** *The family of differential 1-forms*

$$gF^2\,dx = g(x, u)(x^a + u_1 x^{a-2} + \cdots + u_{a-1})^2\,dx,$$

where $g(0,0) \neq 0$, is locally reducible to the form

$$(X^a + U_1 X^{a-2} + \cdots + U_{a-1})^2\,dX$$

by a fibered local change of variables

$$x = x(X, U), \qquad u = u(U).$$

12. **Proof**. In essence, this is the versal deformation theorem for forms of degree 1/2,

$$x^a |dx|^{1/2}.$$

[Similar versal deformation theorems hold for differential forms of arbitrary complex degree p in n variables, $f(x)(dx_1 \wedge \cdots \wedge dx_n)^p$. Their study, which originates from the very problem dealt with here, leads to a theory which is of interest in itself, independent of this problem, and having other applications; for instance, in the theory of Poisson structures in the plane. See [175]–[180].]

In case $n = 1$, $p = 1/k$ (for instance, in our case, where $p = 1/2$), the versal deformation theorem for forms is (in the analytic case) reducible to the usual versal deformation theorem for the function

$$\int x^{ak}\, dx$$

(since the condition 'the derivative of a function is the kth power of a regular function' is a condition that is invariant under diffeomorphisms).

13. We rewrite the constructions in 6 and 7 above in explicit coordinate form. Let q_i be coordinates on L, and p_i the corresponding momenta.

In 6 and 7 we have associated to the hypersurface $H \subset T^*L$ defined by the equation

$$v_0 + v_1 p_1 + \cdots + v_a p_a = 0, \qquad v_i = v_i(q),$$

the field of directions of the vector field on L with components

$$\dot{q}_1 = v_1, \ldots, \dot{q}_a = v_a, \quad \dot{q} = 0 \quad (i > a),$$

and the family of differential 1-forms along its integral curves, which can be written in the form

$$-\frac{v_0}{v_1}\, dq_1 = \cdots = -\frac{v_0}{v_a}\, dq_a.$$

For instance, if $v_1 = 1$, the family of forms is $-v_0\, dq_1$.

Conversely, knowing the components of the field and the 1-form, we can recover the equation of H.

14. Choose coordinates X, U on L according to the lemma in 11. In these coordinates, the field of directions is $\partial/\partial X$, and the family of 1-forms is $F^2(X, U)\, dX$.

Denote the momenta corresponding to these coordinates by p_X and p_U.

The equation of H can be written as

$$F^2(X, U) + p_X = 0,$$

by 13 above.

This formula defines some normal form of the triad, and hence proves the first half of theorem 4. In order to prove the last part, we have to transform the triad to the form described in the theorem.

15. **Lemma**. *The triad*

$$(H : p_X + F^2(X,U) = 0, \quad L : p = 0, \quad l = H \cup L)$$

is symplectically equivalent to the triad

$$\left(H : p_1 + q_1 p_2 + \cdots + q_{a-1} p_a + \frac{q_a^2}{2} = 0, \quad L : p = 0, \quad l : p = q_a = 0 \right).$$

16. In order to prove this lemma, we rectify l (distorting the field). It is convenient to first reduce the leading coefficient of the polynomial F to $1/a!$, and then to reduce the family of forms to $F^2(X,U)\,dX/2$ (by dilations). Then we choose the derivatives of F as new coordinates on L:

$$q_a = F, \; q_{a-1} = \frac{\partial F}{\partial X}, \ldots, \; q_1 = \frac{\partial^{a-1} F}{\partial X^{a-1}}(= X)$$

(we preserve the coordinates q_i on which neither the field of directions nor the 1-form depend).

17. The expressions of the field of directions and the 1-form in the new coordinates are easily determined. Indeed, the derivatives with respect to X of the new coordinates are:

$$\dot{q}_1 = 1, \; \dot{q}_2 = q_1, \ldots, \; \dot{q}_a = q_{a-1}.$$

Along the integral curves of this field of directions,

$$\frac{1}{2} F^2 \, dX = \frac{q_a^2}{2} \, dq_1.$$

Hence, by the formula in 13 above, the equation of the hypersurface H is

$$\frac{q_a^2}{2} + p_1 + q_1 p_2 + \cdots + q_{a-1} p_a = 0.$$

This proves the equivalence of our triad and that in theorem 4. The theorem is thus proved.

Remark. The appearance of F^2 in the normal form of 14 above indicates, it seems, that there is a relation between the variety of polynomials having a root of high multiplicity and the variety of polynomials having many double roots. These two varieties are indeed diffeomorphic (see [8], [104]).

Each polynomial of even degree can be uniquely written as the sum of a square and a polynomial whose degree is (at most) half that of the original polynomial; alternatively, such a polynomial can be uniquely written as the sum of a polynomial with root 0 of multiplicity half the degree and a polynomial whose degree is (at most) half that of the original polynomial.

The normal form in 17 is related to the second decomposition, but in other problems the symplectic structure related to the first decomposition is also useful. Explicit formulas for the diffeomorphism between both Lagrangian varieties (that of the polynomials having roots of high multiplicity and that of the polynomials having many double roots) are given in [8]. Unfortunately, however, they do not explain neither the explicit form of the symplectic structure for which the variety of polynomials having many double roots is Lagrangian, nor the reasons for the diffeomorphism of the swallowtail and the variety of polynomials $x^5 + Ax^3 + Bx^2 + Cx$ having multiple critical values.

The recent experimental discovery of diffeomorphism of the caustic and the Maxwell set of the C_4 singularity (Fig. 91) also indicates the rather unsatisfactory state of the general theory at this point (for instance, generalisations of the latter result to higher C_k singularities have yet to be discovered).

7.5   Contact triads

The foregoing theory has a contact version [166]. Its aim is the study of the Legendre varieties arising in the obstacle problem (that is, of the singularities of the sets of contact elements of the fronts and of the singularities of the set of 1-jets of the function, see § 7.2).

Definition. A *contact triad* (H, L, l) consists of

1. A hypersurface H in a contact manifold that is transversal to the contact hyperplanes;

2. A Legendre submanifold L of this contact manifold;

3. a smooth hypersurface l in L, with H tangent to L of order 1 along l.

We will study the germ of a triad at a point O of l.

Definition. The *Legendre variety generated by the triad* (H, L, l) at O is the image of the germ of l at O under projection of H along its characteristic space.

Example 1. Consider the contact manifold $J^1(M, \mathbf{R})$ of 1-jets of functions on a riemannian manifold without boundary M, equipped with a smooth hypersurface ∂M (we can think of M as being a euclidean space; even the case of the euclidean plane is relevant).

The Hamilton—Jacobi equation $(\nabla u)^2 = 1$, describing the propagation of disturbances along M with velocity 1, defines in $J^1(M, \mathbf{R})$ a hypersurface H. Consider the bundle of geodesics on ∂M that are orthogonal to a fixed hypersurface in ∂M; each geodesic is parametrised by the distance s from this hypersurface. Consider all possible extensions of the function s on ∂M to M. The manifold of 1-jets of all extensions at all points of ∂M is a Legendre submanifold in $J^1(M, \mathbf{R})$.

This submanifold intersects H along the manifold of 1-jets of those extensions whose derivatives along the normal to ∂M vanish. We denote this intersection by l.

Theorem 1. *The triple (H, L, l) generated by a hypersurface ∂M of a riemannian manifold M and a bundle of parametrised geodesics on ∂M is a contact triad.*

Indeed, the order of tangency of H and L along l is 1, since the metric is nondegenerately convex.

Remark. In the notation of Fig. 101, the construction of L from l is symmetric to the construction of γ from l (the symmetry interchanges the maps $\mathcal{I}$ and $\mathcal{II}$; in Fig. 101, l is not on the vertical axis to save space only).

Theorem 2. *The Legendre variety generated by the triad in theorem 1 consists of the contact elements of a front in an obstacle problem (this variety lives in the manifold of oriented contact elements on M).*

Proof. See § 7.2, theorem 5.

In order to describe the normal forms of contact triads, we use the natural contact structures in the odd-dimensional spaces of monic polynomials. We start from the space of binary forms of odd degree, equipped with its natural (SL$_2$-invariant) symplectic structure (see § 1.1).

Projectivisation of this even-dimensional space gives the odd-dimensional projective space of hypersurfaces (i.e. 0-dimensional subvarieties) of fixed odd degree on the projective line.

The skeworthogonal complements to the position vectors form a GL$_2$-invariant field of hyperplanes in the space of nonzero binary forms. This field defines a field of hyperplanes in the projective space of 0-dimensional subvarieties of a fixed degree on the projective line. This field of hyperplanes is a contact structure. This contact structure is natural (is invariant under the action of the group of projective transformations of the line on the space of 0-dimensional subvarieties of a fixed degree).

For an explicit description of this contact structure we use the coordinates (p, q) in the affine part of our projective space of dimension $d = 2a - 1$, consisting of those 0-dimensional subvarieties that do not contain the 'infinite' point $x = \infty$. Such a subvariety can be defined by an equation $f(x) = 0$, where

$$f = e_d + q_2 e_{d-1} + \cdots + q_a e_a - p_a e_{a-1} + \cdots + (-1)^a p_1 \qquad (*)$$

(where $e_d = x^d/d!$ and the signs in front of the p_i alternate). We easily obtain

Theorem 3. *The natural contact structure of the projective space of subvarieties is defined by the 1-form*

$$\alpha = p' \, dq' - q' \, dp' - dp_1,$$

where $p' = (p_2, \ldots, p_a)$, $q' = (q_2, \ldots, q_a)$.

Definition. The *neutral hypersurface* of a 1-parameter group of contact diffeomorphisms is the hypersurface of points at which the velocity vector belongs to the contact plane ($\alpha(v) = 0$, where v is the velocity vector and $\alpha = 0$ is the equation of the contact hyperplane).

Assume that the neutral surface is transversal to the contact hyperplanes. The orbits of the group on the neutral hypersurface are its characteristics.

Shifts of polynomials along the x-axis act on the space of polynomials (*), and preserve its contact structure.

Theorem 4. *The equation of the neutral hypersurface is $k = 0$, where*

$$k = p_2 + p_3 q_2 + \cdots + p_a q_{a-1} + \frac{q_a^2}{2}.$$

Indeed, the components of the velocity field of the group of shifts along the x-axis are:

$$\dot{q_i} = q_{i-1}, \quad \dot{p_i} = -p_{i+1} \qquad (\text{where } p_{a+1} \equiv q_a, q_1 \equiv 1).$$

Now we are ready to give the normal form of a contact triad. It is defined by the neutral hypersurface $k = 0$ in the contact manifold of polynomials (*) and by the Legendre manifold of polynomials having at the origin a root of multiplicity at least a.

Theorem 5. *The triple*

$$(H : k = 0, \quad L : p = 0, \quad l : p = q_a = 0)$$

is a contact triad in the space $\mathbf{R}^{2a+1}$ equipped with th contact structure

$$\alpha = p'\, dq' - q'\, dp' - dp_1.$$

The proof is an easy verification, based on the expression for k (given in theorem 4).

Definition. The contact triad in theorem 5 and the Legendre variety defined by it are called the *standard contact triad* in $\mathbf{R}^{2a-1}$ and the *standard singular Legendre variety*. The dimension of this standard variety is $m = a - 2$. We will denote the standard singular Legendre variety by Σ^m.

Theorem 5 implies

Corollary 6. *The standard singular Legendre variety Σ^{a-2} consists of those polynomials (*) for which*

$$q_2 = 0, \quad p_2 = -p_4 q_3 + \cdots + p_a q_{a-1} + \frac{q_a^2}{2}$$

and which have a root of multiplicity at least a. The contact structure of this space is defined by the 1-form

$$\alpha = p'' \, dq'' - q'' \, dp'' - dp_1,$$

where $q'' = (q_3, \ldots, q_a)$, $p'' = (p_3, \ldots, p_a)$.

Combining this result with those in § 7.3, concerning polynomials with roots of high multiplicity, we find

Corollary 7. *The standard singular Legendre variety Σ^m is diffeomorphic to the unfurled swallowtail of dimension m, that is, to the variety of polynomials of the form*

$$x^{2m+1} + u_1 x^{2m-1} + \cdots + u_{2m} = (x - t)^{m+1}(x^m + \cdots).$$

Example. For $a = 3$ corollary 6 provides the polynomials

$$\frac{1}{120} x^5 + \frac{q_3}{6} x^3 - \frac{p_3}{2} x^2 - p_1,$$

$$\alpha = p_3 \, dq_3 - q_3 \, dp_3 - dp_1.$$

The polynomials of this form and having a root of multiplicity at least 3 form a Legendre curve

$$\Sigma^1 = \left\{ f = \frac{1}{120} (x - t)^4 (x + 4t) \right\},$$

that is,

$$q_3 = -\frac{t^2}{2}, \quad p_3 = -\frac{t^3}{3}, \quad p_1 = -\frac{t^5}{30}.$$

This curve is diffeomorphic to a semicubic parabola. Its singularity is responsible for the 5/2-singularities of evolvents of planar curves at generic points of inflection of the tangents.

Remark. The variety Σ^{a-2} is quasihomogeneous. We assign weights to the coordinates:

$$\deg q_i = i - 1, \qquad \deg p_i = 2a - i.$$

Then $\deg k = 2a - 2$, $\deg \alpha = 2a - 1$. The variety Σ^{a-2} is the image of the space generated by the coordinates of weights $(1, \ldots, a - 2)$ under a homogeneous map.

Example. For $a = 4$ the coordinates $(q_3, q_4, p_4, p_3, p_1)$ have weights $(2, 3, 4, 5, 7)$. The Legendre variety Σ^2 is the image of the (t, s)-plane, where

$$\deg t = 1, \qquad \deg s = 2,$$

since

$$\Sigma^2 = \left\{ f = \frac{1}{7!} (x - t)^5 (x^2 + 5tx + s) \right\}.$$

This variety is diffeomorphic to the variety of monic polynomials of degree 6, with trace 0, and a root of multiplicity at least 4.

Definition. The *suspension* over the standard triad in $\mathbf{R}^{2a-1}$ is the triad in $\mathbf{R}^{2b-1}$ equipped with the contact structure

$$\alpha = p'\, dq' - q'\, dp' - dp_1, \qquad p' = (p_2, \ldots p_b), \quad q' = (q_2, \ldots, q_b),$$

defined by

$$\left(H : p_2 + p_3 q_2 + \cdots + p_a q_{a-1} + \frac{q_a^2}{2} = 0, \quad L : p = 0, \quad l : p = 0, q_a = 0 \quad (b \geq a) \right).$$

Theorem 8. *The germs at the origin of the standard triads (and the germs of their suspensions) are stable (as germs of contact triads, considered up to contact equivalence).*

A generic triad is locally, in a neighborhood of each point, contactomorphic to a standard triad or its suspension.

Corollary 9. *The Legendre varieties formed by the contact elements tangent to a front in a generic obstacle problem are stable with respect to contactomorphisms, and are locally, in a neighborhood of each point, contactomorphic to the standard Legendre varieties (or their suspensions); hence they are locally diffeomorphic to the unfurled swallowtails.*

In order to prove theorem 8 it suffices to contactify theorem 4 in § 7.4. Namely, choose a 1-form α ,defining the contact structure, for which the kernel of the form $d\alpha$ at the points of the hypersurface which is the first term in the triad, is tangent to this hypersurface.

Projection of the contact manifold (along the integral curves of the field of kernels) defines a symplectic structure in the even-dimensional space of integral curves (the image of $d\alpha$).

The projection of a contact triad is then a symplectic triad. The initial contact triad can be recovered from its symplectic projection by contactifying (by adding the value of the generating function of the Lagrangian manifold as a new coordinate). Applying this general construction to the standard symplectic triad (in theorem 4 in § 7.4) gives the standard contact triads in theorem 8. Now we are ready to reduce to normal form the singularities of the 1-graph of the time function (this 1-graph is the variety of 1-jets of the multivalued time function, and it lives in $J^1(M, \mathbf{R})$).

The 1-graph of the time function (denoted by Γ in § 7.2, Fig. 101) is formed by the characteristics of the hypersurface H (denoted by $SJ^1(M, \mathbf{R})$ in § 7.2) that intersect the manifold l of the triad (H, L, l) in theorem 1 (see theorem 10 in § 7.2).

Consider the obstacle problem in which the obstacle is bounded by a generic hypersurface ∂M in a riemannian manifold M. Consider on ∂M a generic bundle of geodesics, parametrised by the distance s from a given hypersurface in ∂M. This bundle defines the time function on M (it coincides with s on ∂M).

Theorem 10. *The variety Γ of 1-jets of the time function is locally contactomorphic to the variety of polynomials (*) having a root of multiplicity exceeding a.*

The 1-jets of the time function at points of a single ray (that is, on a single geodesic of M) are sent to the polynomials (*) by the diffeomorphism above; these polynomials are obtainable one from another by shift along the x-axis.

Proof. The triple (H, L, l) generated by s is a contact triad, by theorem 1. This triad is generic, provided that the obstacle and bundle are generic. Hence it is locally equivalent to the triad in theorem 5 (by theorem 8; we do not need suspensions, since we may regard as normal form a germ of a standard triad at an arbitrary point).

For a triad as in theorem 5 the manifold l consists of the polynomials with root at 0 of multiplicity exceeding a. The characteristics of the hypersurface H in theorem 5 are the orbits of shifts along the x-axis (by theorem 4). The orbits of the points of l sweep the variety of polynomials (*) having a root of multiplicity exceeding a, as required.

Finally we will give explicit formulas for the singularities of the varieties of contact elements of a front and the 1-graphs of the time function, as well as of the subvarieties of the space of polynomials of odd degree,

$$x^n + a_1 x^{n-1} + \cdots + a_n, \qquad n = 2k + 1.$$

The natural (SL_2-invariant) contact structure of this space of polynomials is defined by the equation

$$\sum i!j!(-1)^i a_i \, da_j = 0 \qquad \text{(where } i + j = n, \ a_0 = 1\text{)}.$$

Theorems 10 and 8 imply

Corollary 11. *The 1-graph (a variety of 1-jets) of a generic time function in a generic obstacle problem in an m-space is locally contactomorphic to the singular Legendre variety of polynomials*

$$(x - t)^{m+2}(x^{m-1} + \cdots)$$

in the space of polynomials of degree $n = 2m + 1$, equipped with its natural contact structure. The restriction of the time function to this subvariety has the form $\pm a_1 + \mathrm{const}$; points on a single ray correspond to shifts along the x-axis of a single polynomial.

Corollary 12. *The variety of contact elements of a generic front in a generic obstacle problem in an m-space is locally contactomorphic to the singular Legendre variety of* *polynomials*

$$x^n + x^{n-1} + a_{n-2}x^{n-2} + \cdots + a_0 = (x - t)^{m+1}(x^{m-1} + \cdots)$$

in the space of polynomials of degree $2m$ with $a_{n-1} = 1$, equipped with the natural contact structure

$$\sum i!j!(-1)^i(1+j)a_i\, da_j = 0 \qquad (\text{where } 0 < i < n,\ i + j = n - 1).$$

Remark. In [166], [170] the corresponding formulas are written out wrongly.

The structure of generic fronts themselves and of the graphs of the time function in higher-dimensional obstacle problems is unknown, even topologically.

For instance, it is unknown whether in a generic obstacle problem the shortest time is topologically equivalent to a Morse function. This does hold in the corresponding finite-dimensional problem: topologically, the maximum and minimax functions

$$f(y) = \max_x F(x, y), \qquad f(z) = \min_y \max_x F(x, y, z)$$

are Morse functions, provided that the families F are generic families of functions on compact manifolds (Matov [181]).

It is also unknown whether in the obstacle problem the general principle of fragility of good things, according to which at a singular point of the boundary of the set of 'good' objects the directions of bad transformation make up more than one half of the space of directions, holds. It is clear, however, that in the obstacle problem the 'good' objects are the inaccessible ones.

7.6 Hypericosahedral singularity

The results in this section are due to O.P. Shcherbak [6], whose posthumous paper [144] contains all proofs. The *hypericosahedron* is the regular 600-hedron in euclidean 4-space. In order to describe its 120 vertices, we begin with the ordinary icosahedron rotation group. This group, having 60 elements, is a subgroup of SO(3). The double covering $SU(2) \to SO(3)$ sends the icosahedron rotation group to a subgroup of SU(2), called the *icosahedron binary group*. Its 120 elements live in $SU(2) \approx S^3$. Their convex hull is the hypericosahedron.

The hypericosahedron symmetry group is generated by 4 reflections of euclidean space. This reflection group is denoted by H_4. The corresponding Dynkin diagram is $\bullet \overset{5}{-} \bullet - \bullet - \bullet$, where 5 stands for an angle of $\pi - \pi/5$. The group H_4 contains 120^2 elements. It acts on the vertices of the hypericosahedron as left and right multiplication by elements of the binary group (hence it is isomorphic to the direct product of the icosahedron binary group with itself).

The Coxeter group H_4 is not a crystallographic group: it does not preserve any lattice in euclidean 4-space. However, it is related to the lattice E_8, from which it can be constructed as follows.

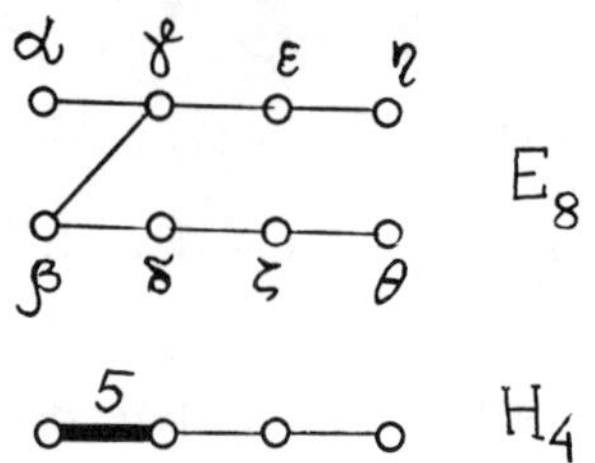

Figure 109: Construction of the group H_4 from the lattice E_8

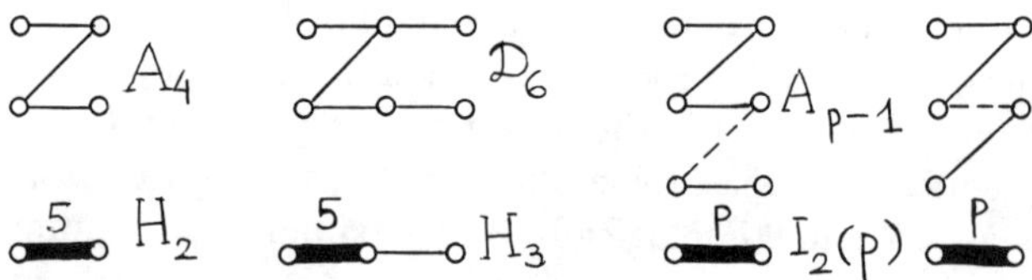

Figure 110: Foldings of the Dynkin diagrams generating the groups H_2, H_3, $I_2(p)$

Consider the Dynkin diagram E_8, folded as in Fig. 109. Each point represents a mirror, and this gives rise to a reflection of euclidean 8-space. Two reflections α, β commute. Hence the product $\alpha\beta$ is a well-defined orthogonal transformation of the 8-space (preserving the lattice E_8). The 4 products $\alpha\beta$, $\gamma\delta$, $\epsilon\zeta$, $\eta\theta$ generate a group of orthogonal transformations preserving the lattice E_8. It is easily seen that this group is isomorphic to H_4.

Hence the folding in Fig. 109 determines a group representation of H_4 in euclidean 8-space (preserving the lattice E_8). This representation is reducible. Namely, 8-space is the orthogonal sum of two 4-dimensional irreducible representation spaces of H_4. Thus, the group H_4 of reflections is the group generated by the 4 products above, acting in the 4-dimensional (irrational) space of the irreducible representation.

This relation between H_4 and E_8 is crucial in the study of the discriminant of H_4 (which is an algebraic hypersurface in the orbit space of H_4).

Remark. The other noncrystallographic Coxeter groups can be defined by a similar construction from foldings of the other Dynkin diagrams, shown in Fig. 110: $A_4 \to H_2$, $D_6 \to H_3$, $A_{p-1} \to I_2(p)$.

We will now describe the discriminant of H_4.

Theorem 1 (see [144]). *The discriminant of the reflection group H_4 admits a parametrisation*

$$t_2 = x,$$
$$t_{12} = xz + \frac{y^2}{2},$$

$$t_{20} \quad = xy^3 + \frac{z^2}{2},$$

$$t_{30} \quad = xy^3 z + \frac{y^5}{5} + \frac{z^3}{3},$$

where t_k are the basic invariants (k being the degree) and (x, y, z) are the parameters of weights 2, 6, 10.

In Fig. 111 the discriminant of H_4 is presented as the perestroika of its 2-dimensional sections (its sections by 3 parallel 3-dimensional spaces of generic direction are shown).

Theorem 2 ([144]). *The bifurcation diagram of zeros of the family of functions*

$$f(x, y) = \frac{x^5}{5} + ax^3(s_2 y + s_{12}) + x(s_2 y + s_{12})^2 + y^3 + s_{20} y + s_{30},$$

depending on 4 parameters $(s_2, s_{12}, s_{20}, s_{30}))$ (a being a nonzero constant), has two components, each diffeomorphic to the discriminant of H_4.
  If $a = 1$ or $2/3$, these two hypersurfaces coincide.
  The map $(\mathbf{C}^4, 0) \to (\mathbf{C}^8, 0)$, inducing the family from the versal deformation E_8, is a (local) embedding.
  If $a = 2/3$ it maps the bifurcation diagram of the family onto the stratum $4A_2$ of the discriminant of E_8 (4 nonMorse points), which is, hence, diffeomorphic to the discriminant of H_4.
  If $a = 1$ this map sends the bifurcation diagram of the family onto the stratum $(A_1 A_1)(A_1 A_1)(A_1 A_1)(A_1 A_1)$ of the discriminant of E_8 (4 pairs of equal critical values), which is hence also diffeomorphic to the discriminant of H_4.

Remark. Similar results hold for H_2 and H_3. The corresponding families

$$\frac{x^5}{5} + as_2 x^3 + s_2^2 x + y^2 + s_{10}, \qquad a \neq 0,$$

$$\frac{x^5}{5} + ax^3 y + xy^2 + s_2 y^2 + s_6 y + s_{10}, \qquad a^2 \neq \frac{4}{5},$$

are induced from the versal deformations of A_4 and D_6. The case $a = 2/3$ corresponds to the strata $2A_2$ in the discriminant of A_4, and to $3A_2$ in the discriminant of D_6. The case $a = 1$ corresponds to the strata $(A_1 A_1)(A_1 A_1)$ in the discriminant of A_4, and to $(A_1 A_1)(A_1 A_1)(A_1 A_1)$ in the discriminant of D_6.

The discriminants of the irreducible crystallographic Coxeter groups are the fronts of the corresponding Legendre singularities (see § 3.2). This relation was extended to the noncrystallographic groups by A.B. Givental [8]. We start from the following generalisation of the Ljashko—Looijenga theory.

Consider a holomorphic vector field in the orbit space $\mathbf{C}^k$ of an irreducible Coxeter group. The discriminant is a singular hypersurface in this space. Its tangent plane

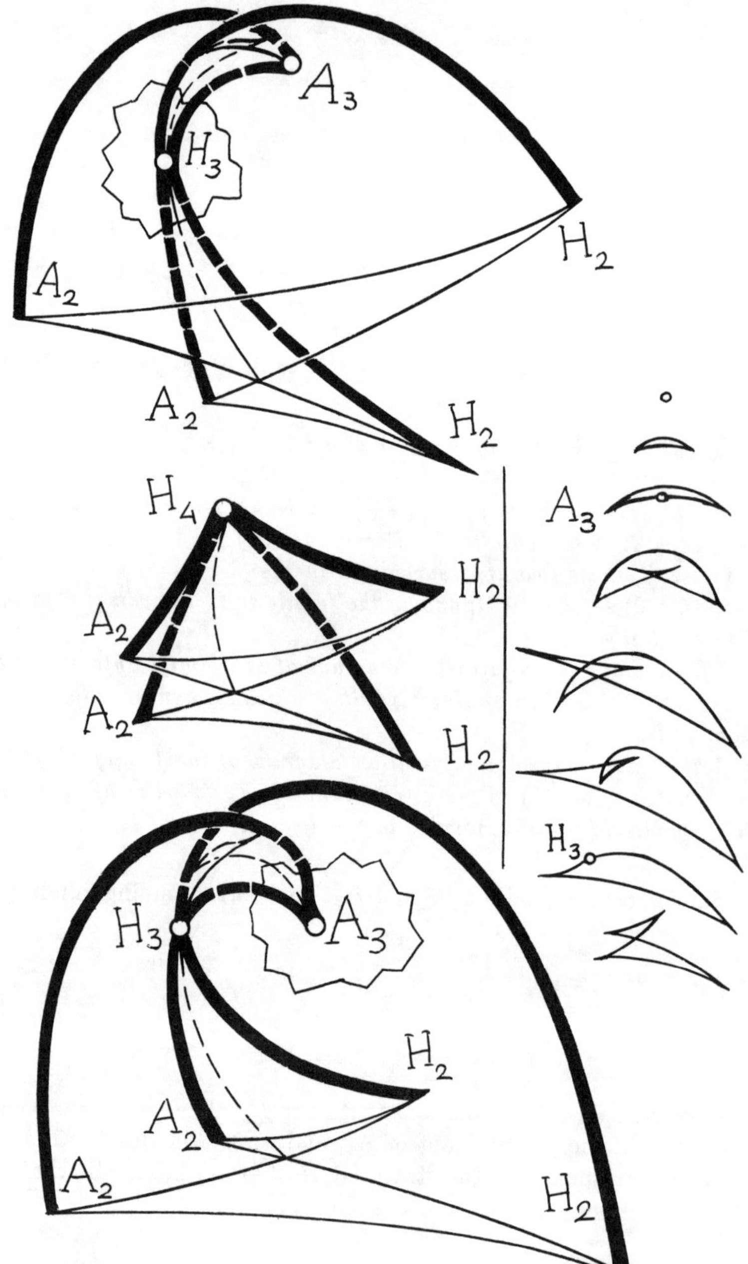

Figure 111: The discriminant of H_4 as the perestroika of a surface

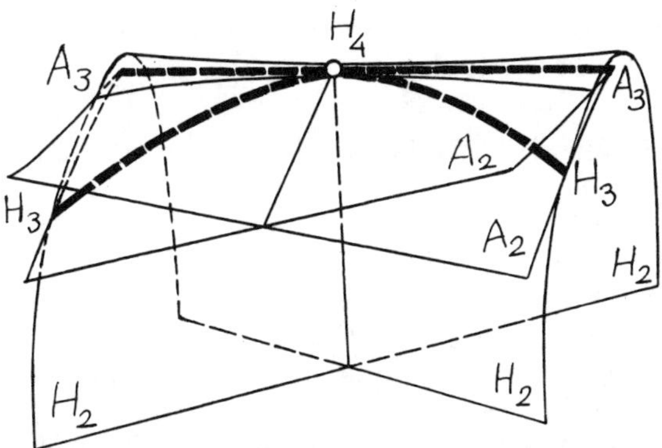

Figure 112: The caustic of H_4

at the origin is defined by the equation $dt_h = 0$, where t_h is the basic invariant of highest degree. This degree h is called the *Coxeter number*.

Theorem 3 (O.V. Ljashko [159], A.B. Givental [8]). *All holomorphic vector fields that are transversal to the discriminant at the origin are locally holomorphically equivalent (i.e. one can be transformed into the other by a local diffeomorphism preserving the discriminant hypersurface).*

Consider now the local fibration $(\mathbf{C}^k, 0) \twoheadrightarrow (\mathbf{C}^{k-1}, 0)$ of the orbit space into the trajectories of a generic holomorphic vector field (transversal to the discriminant). The trajectories that nongenerically intersect the discriminant (the intersection consists of less than k points) form a hypersurface in the space of trajectories $\mathbf{C}^{k-1}$. This hypersurface is called the *bifurcation diagram of functions* (see § 5.3 for motivation).

Theorem 4 (O.V. Ljashko [159], A.B. Givental [8]). *The higher homotopy groups of the complement of the bifurcation diagram of functions of an irreducible Coxeter group W are trivial, while the fundamental group is a subgroup of finite index in the braid group on k strings. This index equals*

$$\frac{k!h^k}{|W|}$$

(where $|W|$ denotes the number of elements of W).

The caustic of the singularity H_4 (which is the part of the bifurcation diagram of functions corresponding to the cuspidal edge of the discriminant hypersurface) is shown in Fig. 112. In $\mathbf{C}^3$ this surface consists of an ordinary umbrella H_2 and a folded umbrella A_2 that are tangent to each other of order 3 along the line H_3.

We will now regard the discriminant as the graph of a 'multivalued function' on $\mathbf{C}^{k-1}$. The set of local 'differentials' of this 'function' form a Lagrangian variety in

$T^*\mathbf{C}^{k-1}$. The corresponding Legendre variety is formed by the contact elements of $\mathbf{C}^k$ that are tangent to the discriminant.

Theorem 5 (A.B. Givental). *The Lagrangian projection of the Lagrangian variety corresponding to a Coxeter group is simple. The corresponding Lagrangian variety is diffeomorphic to the product of the curve $p^2 = q^r$ and a smooth manifold, with r as in the following table:*

group	A_k, D_k, E_k	B_k, C_k, F_4	H_k	$I_2(m)$
r	1	2	3	$m - 2$

Conversely, any simple Lagrangian projection of a Lagrangian variety that is diffeomorphic to a product of planar curves is locally equivalent to a projection generated by a Coxeter group (or to its trivial suspension).

Remark. Here the word 'simple' means that the Lagrangian projection $L \hookrightarrow T^*B \twoheadrightarrow B$, where L is a *fixed* variety, has no moduli (all Lagrangian projections of a fixed variety and sufficiently close to the given projection are locally, in a neighborhood of an arbitrary point, equivalent to a projection belonging to some finite list)[1].

Example. The Lagrangian variety corresponding to H_k has an ordinary cuspidal edge ($r = 3$). Hence the only possible singularity of the Legendre variety formed by the contact elements tangent to the discriminant of H_4 is a semicubical cuspidal edge. The complicated singularity in Fig. 111 reflects the nongenericity of the projection of this Legendre 4-variety from the 7-space of contact elements to the 4-space in which the discriminant lives.

Thus we are led to the problem of classifying the Legendre projections of Legendre varieties with a semicubical cuspidal edge. We start with the simplest case, in which a Legendre curve with an ordinary cusp is projected to the plane along the fibers of a Legendre fibration.

Theorem 6. *A Legendre curve with a semicubical cusp is projected onto a curve with an H_2 singularity (of order $5/2$), provided that the tangent of the curve at the cusp is not vertical (i.e. does not coincide with the direction of the fiber). Such a projection is Legendre equivalent to the projection of the set of contact elements that are tangent to the curve $x^5 = y^2$, onto this curve.*

Remark. It follows that the only singularities of the Legendre projection of a generic Legendre curve with ordinary cusps are cusps of order $5/2$ (the images of the cusps of the Legendre curve) and of order $3/2$ (the images of the points at which the curve is smooth but vertical).

This remark implies that a curve which is projectively dual to a generic planar curve with a cusp singularity of order $5/2$ has a singularity of the same type. This

---

[1] The definition of 'simple' in [8] is slightly different, but the result is valid for both definitions.

is not true for certain nongeneric curves; for instance, for the 'normal form' defined by the equation $y = x^{5/2}$ in affine coordinates. Indeed, the corresponding Legendre curve is tangent to the fiber of the second canonical Legendre fibration $PT^*P^2 \twoheadrightarrow P^{2*}$.

The curve $y = x^2 + x^{5/2}$ is generic with respect to both canonical fibrations $PT^*P \twoheadrightarrow P^2$ and $PT^*P \twoheadrightarrow P^{2*}$, and its dual has an H_2 singularity of order $5/2$, which is diffeomorphic to the singularity at the origin of the initial curve.

We will now study Legendre projections of Legendre surfaces with semicubical cuspidal edges. The edge itself is a smooth integral curve of the contact structure. A generic integral curve of a contact structure is nowhere vertical. Hence its projection is a smooth line that locally can be transformed into a straight line, using a diffeomorphism of the base 3-space. The ambient contact manifold and its Legendre projection can be identified with the space of contact elements of the base 3-space. The elements corresponding to the cuspidal edge can be made parallel, using a new diffeomorphism preserving the projection line.

Generically, at a point of the cuspidal edge the tangent plane of the Legendre variety is nonvertical (does not intersect the tangent plane of the fiber). At certain points of the cuspidal edge, however, it may become vertical (for a generic Legendre surface with a cuspidal edge such points are isolated and the corresponding map from the edge to the Lagrangian Grassmann manifold of the hyperplane of the contact structure is transversal to the train[2] of the Lagrangian plane that is tangent to the fiber). This generates the singularity H_3.

Consider a generic fibration into planes of the base 3-space. The front of our Legendre variety intersects these fibers along planar curves. The contact elements of a fiber containing the tangent directions of the intersection curve form a Legendre variety in the 3-dimensional contact space of contact elements of the plane. This Legendre variety is the Legendre boundary of the initial Legendre surface. It has a semicubical cusp at the point corresponding to the cuspidal edge of the initial surface.

Hence we have constructed a 1-parameter family of Legendre curves in the spaces of Legendre fibrations of contact elements of planes. At the cusp point of any curve the tangent line belongs to the contact plane at that point. For generic parameter values the tangent line is nonvertical (does not coincide with the direction of the fiber). However, for certain parameter values the tangent becomes vertical. This is the case for the points H_3, here described.

We locally identify our 3-space of the Legendre fibration with the space of 1- jets of functions of 1 variable,

$$J^1(\mathbf{R}, \mathbf{R}) \twoheadrightarrow J^0(\mathbf{R}, \mathbf{R}), \qquad (x, y, p) \mapsto (x, y).$$

Since $dy = p\,dx$ along Legendre curves, we can reconstruct these curves from their Lagrangian projections $\{(x, p)\}$. These Lagrangian curves have semicubical cusps at the origin. Generically, the tangent at the cusp is nonvertical, but it becomes vertical

---

[2] The *train* of a Lagrangian vector space is the variety of Lagrangian vector spaces that are not transversal to it, that is, whose intersections with the given Lagrangian vector space contain nonzero vectors.

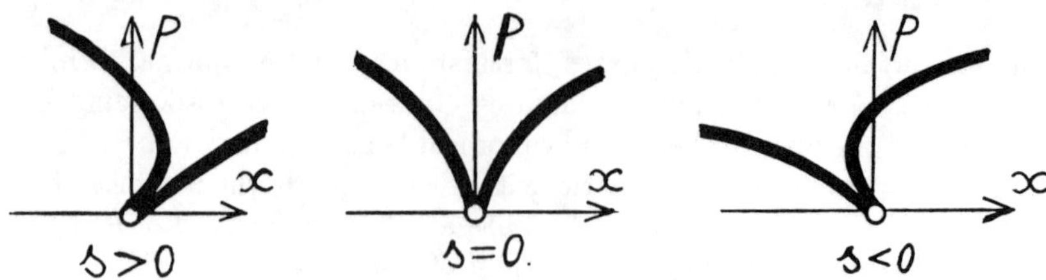

Figure 113: The H_3 perestroika of Lagrangian curves with cusps

$(dx = 0)$ for the parameter value corresponding to the singularity H_3. The perestroika of the Lagrangian curve is shown in Fig. 113.

Example. $p = t^2$, $x = t^3 + st^2$, where s is the parameter. The corresponding Legendre curve is

$$p = t^2, \quad x = t^3 + st^2, \quad y = \int p\,dx = \frac{3}{5}t^5 + \frac{2}{3}st^3.$$

The front is the curve in the (x, y)-plane described parametrically by

$$x = t^3 + st^2, \quad y = \frac{3}{5}t^5 + \frac{2}{3}st^3.$$

For $s = 0$ the front has a singularity H_2 of order 5/2 at the origin, where the Legendre curve has a cusp, and a singularity A_2 of order 3/2 at the point at which the Legendre curve is smooth but vertical $(3t + 2s = 0)$. The perestroika of the front is shown in Fig. 5. The surface in 3-dimensional 'space–time' $\{(x, y, s)\}$ swept by the fronts is diffeomorphic to the discriminant of H_3 (Fig. 102). It is also the front of the corresponding Legendre surface with a cuspidal edge.

This example provides a normal form of the generic Legendre projection of a surface with a cuspidal edge, of the singularity of its front, of its decomposition into level lines of a generic function, and of the generic perestroikas of Legendre projections of Legendre curves with semicubical cusps.

Warning. This example does *not* provide a normal form for the perestroika of the projection of a *Lagrangian* curve, but just an example of the latter. Indeed, consider two fibers $x = \text{const}$ that are not transversal to the Lagrangian curve. Together with the curve they bound two areas, $(u(s), v(s))$. The resulting curve in the (u, v)-plane is intrinsically related to the Lagrangian perestroika. It varies, however, from one family to another. Hence the Lagrangian perestroika of H_3 has functional moduli.

Remark 1. The discriminant of H_3 is locally diffeomorphic to the union of the tangent lines of the space curve

$$x = s, \quad y = s^3, \quad z = s^5$$

(O.V. Ljashko [4]), or of any curve

$$x = s + \cdots, \quad y = s^3 + \cdots, \quad z = s^5 + \cdots,$$

where the dots indicate higher-order terms (O.P. Shcherbak [5]).

Remark 2. Consider in projective space a surface with a singularity H_3. Generically, the dual surface has a singularity H_2. However, it can have a singularity H_3. This event has codimension 1 (in projective space it happens generically for certain parameter values in 1-parameter families of surfaces with singularity H_3).

The discriminant of H_4, as well as those of H_3 and H_2, can also be described as the singularity of a generic front of a Legendre variety with a semicubical cuspidal edge (of dimension 2 in the case of H_4).

The cuspidal edge is a 2-dimensional integral surface of the contact structure in the 7-dimensional ambient manifold. Generically, at isolated points such surfaces are vertical (are tangent to a fiber of a fixed Legendre fibration).

Indeed, the projection of the cuspidal edge to the space of the corresponding 6-space of the Legendre fibration is an isotropic 2-surface. The Grassmann manifold of isotropic 2-planes in the symplectic 6-space is 7-dimensional. The train of a fixed Lagrangian subspace (formed by those isotropic 2-planes not transversal to it) is 5-dimensional. The codimension of the train is 2. The tangent planes of the cuspidal edge are parametrised by 2 parameters. Hence the plane becomes (transversally) vertical at certain isolated points of the cuspidal edge (here we use a transversality theorem, based on the surjectivity of the 'Gauss' map sending an isotropic submanifold, with a distinguished point, to the tangent space at this point, translated to the origin of the ambient euclidean symplectic space).

Thus, the cuspidal edge of a generic Legendre subvariety of dimension 3 is vertical at certain isolated singular (H_4) points.

The Legendre projection of the cuspidal edge is not smooth at these points. This 2-surface in 4-space can be parametrised by

$$t_2 = a, \quad t_{12} = ac, \quad t_{20} = \frac{1}{2}c^2, \quad t_{30} = \frac{1}{3}c^3.$$

Along this surface the front has a cuspidal edge (H_2) of order 5/2. The singularity of the H_2 stratum of the H_4 front is thus the open umbrella (see § 5.6).

Remark. Thus, the H_2 stratum in the 4-dimensional orbit space of the reflection group H_4 is diffeomorphic to the conormal bundle of a semicubic parabola. It would be interesting to have an understanding of the origin of the corresponding symplectic structure of the orbit space of the singularity H_4 (is the fact that the open umbrellas live in spaces whose dimensions are multiples of 4 related to hyperkählerian structures?).

Let us return to the Legendre variety of dimension 3 with cuspidal edge H_2 of dimension 2. The tangent plane to this variety at the points of the cuspidal edge may become vertical (tangent to the fiber). Generically, this happens at the points of a curve. The corresponding singularity of the front is H_3 (there the front is locally diffeomorphic to the product of the discriminant of H_3 and a straight line). This H_3 curve on the cuspidal edge contains the singular point of type H_4, and is smooth at that point. The projection of this smooth curve remains smooth: it is the H_3 stratum of the discriminant of H_4 (it coincides with the t_{12}-axis in the coordinates above).

The discriminant of H_4 should be described as a perestroika in a 1-parameter family of H_3 singularities (like the discriminant of H_3 is described as a perestroika in a 1-parameter family of H_2 singularities). This description is left to the reader.

Remark. Another description of the discriminant of H_4 is due to O.P. Shcherbak: consider the family of curves

$$x^5 + y^3 + u_2 x^3 y + u_{12} x^3 + u_{20} y + u_{30} = 0.$$

The parameter values u corresponding to singular curves form a hypersurface in 4-space. This hypersurface has 2 irreducible components. One of these is diffeomorphic to the discriminant of H_4.

The formula above describes an embedding of the local algebra D_4 in the local algebra E_8, inducing on the former the same grading as that given by convolution of invariants of H_4. This remark, contained in [98] (remark 7, § 9) was the starting point of Shcherbak's work on H_4.

7.7   Normal forms of singularities in the obstacle problem

We define a family of 'length functions' on the surface of the obstacle. A function in this family will depend, besides on a parameter, on a point of the ambient manifold. The value of a function will be the sum of the distances from an initial 'source' submanifold (along the geodesics of the obstacle) and the distance from the 'goal' point of the ambient manifold (along its geodesics; we will assume that both summands are smooth). The main observation in Shcherbak's theory of the obstacle problem is

Lemma. *The multiplicities of the critical points of a length function are even.*

The proof is obvious from Fig. 114: the simplest singularity of a length function of 1 variable is the cubical (A_2) singularity, since the function is monotone.

The multiplicity of a more complicated singularity of a length function is twice the number of simplest (A_2) singularities into which it bifurcates under a slight variation (taking into account complex singularities, of course).

In the versal deformation spaces of functions with simple critical points, O.P.

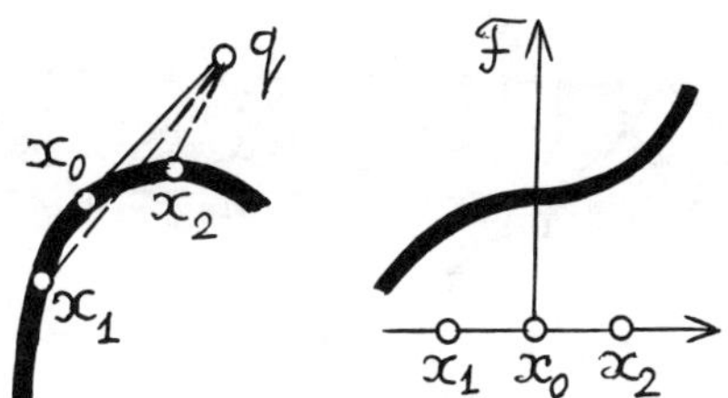

Figure 114: The graph of a length function F

Shcherbak has found the (maximal) strata of those functions whose critical points have even multiplicities.

Theorem 1 ([144]). *Any maximal nonsingular subfamily of R^+-versal deformations of a simple function germ is equivalent (in the R^+-sense) to one of the following families in Shcherbak's list:*

$$
\begin{aligned}
\Xi_k(\subset A_{2k}) \quad & \int_0^x (u^k + q_1 u^{k-2} + \cdots + q_{k-1})^2 \, du, \\
\Omega_k(\subset D_{2k}) \quad & \int_0^y (u^{k-1} + q_1 u^{k-3} + \cdots + q_{k-3}u + x)^2 \, du + q_{k-2}x^2 + q_{k-1}x, \\
A_3(\subset E_6) \quad & x^3 + y^4 + q_1 y^2 + q_2 y, \\
A_4(\subset E_8) \quad & x^3 + y^5 + q_1 y^3 + q_2 y^2 + q_3 y, \\
H_4(\subset E_8) \quad & x^3 + \int_0^y (u^2 + q_1 x + q_2)^2 \, du + q_3 x.
\end{aligned}
$$

Remark. The deformation of D_{2k} contains one more maximal subfamily (equivalent to that in the list). The deformations of E_6 and E_8 also contain singular subfamilies (which are absent in the other cases).

We will now consider the Legendre maps corresponding to the families above. A family $F(x;q)$ defines the Lagrangian subvariety

$$
\left\{ (p,q) : \exists x : \frac{\partial F}{\partial x} = 0, p = \frac{\partial F}{\partial q} \right\}
$$

in the space of the cotangent bundle $T^* \mathbf{R}^{k-1} \twoheadrightarrow \mathbf{R}^{k-1}$, $(p,q) \mapsto q$.

Theorem 2 (A.B. Givental [8]). *The following families in Shcherbak's list generate simple Lagrangian maps:*

$$
\Xi_k \quad (k \geq 3), \quad \Omega_k \quad (k \geq 4), \quad H_k \quad (k = 2,3,4; \quad H_2 = \Xi_2, H_3 = \Omega_3).
$$

The corresponding Lagrangian varieties are products of d-dimensional unfurled swallowtails with smooth manifolds, with d as in the following table:

family	Ξ_k	Ω_k	H_k
d	$k-1$	$k-2$	1

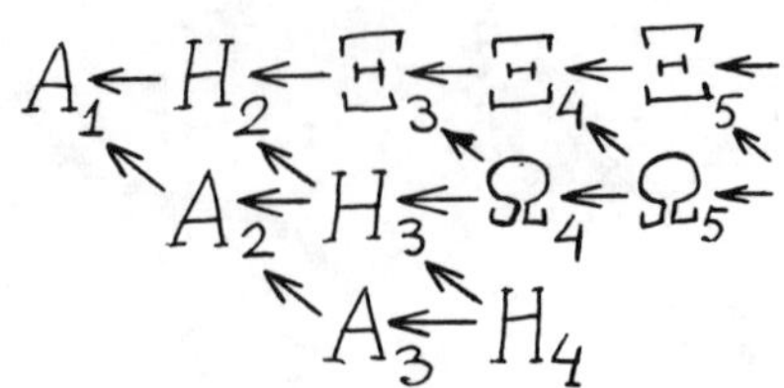

Figure 115: Adjacencies of simple Lagrangian projections of unfurled swallowtails

A simple Lagrangian map from the product of an unfurled swallowtail of arbitrary dimension and a smooth manifold is locally equivalent to one of the maps Ξ_k, Ω_k, H_k (or to the trivial suspension of one of these).

Remark. The adjacencies of these singularities are shown in Fig. 115. The line (H_2, H_3, H_4) corresponds to the 1-dimensional unfurled swallowtail, that is, to the semicubic parabola. The corresponding Lagrangian variety has a semicubical cuspidal edge. The next line (Ξ_3, Ω_4) corresponds to the 2-dimensional unfurled swallowtail singularity of the Lagrangian variety. The left line (A_1, A_2, A_3) corresponds to the 0-dimensional unfurled swallowtail, and hence to smooth Lagrangian varieties. These projections have the usual Lagrangian singularities of § 1.3.

The normal forms Ξ and Ω in theorem 2 can be presented in slightly modified form, using the natural symplectic and contact structures in spaces of polynomials.

Case Ξ_{n+1}. We start with the symplectic space of polynomials (§ 1.1)

$$x^{2n+1} + a_1 x^{2n-1} + \cdots + a_{2n}.$$

The polynomials having a root of multiplicity exceeding n form an unfurled swallowtail of dimension n, which is a Lagrangian variety.

Consider the fibration

$$(a_1, \ldots, a_{2n}) \mapsto (a_1, \ldots, a_n).$$

This gives a Lagrangian projection from the unfurled swallowtail.

Theorem 3. *Any generic Lagrangian projection is locally reducible to the one above, by a local symplectomorphism preserving the unfurled swallowtail (in a neighborhood of the vertex of the swallowtail).*

For instance, the projection Ξ_{n+1} in theorem 2 is reducible to this normal form.

The corresponding caustic (in this case formed by the projections of the singular points of the unfurled swallowtails, since it is nowhere tangent to the fibers) is the ordinary swallowtail of dimension $n - 1$ (since the projection above is equivalent to a repeated differentiation of polynomials).

Case Ω_{n+1}. In order to describe the singularity of Ω_{n+1}, consider the product of the unfurled swallowtail of dimension $n - 1$ and a line. We will call it a *cylinder*. This n-dimensional cylinder is a Lagrangian subvariety in the space of polynomials

$$x^{2n} + a_1 x^{2n-1} + \cdots + a_{2n}, \tag{1}$$

equipped with the usual symplectic structure of § 1.1. The required Lagrangian subvariety is formed by the polynomials having a root of multiplicity exceeding n.

Consider the Lagrangian fibration

$$(a_1, \ldots, a_{2n}) \mapsto (a_1, \ldots, a_{n-1}; a_{n+1}).$$

This gives a Lagrangian projection from the cylinder.

Theorem 4. *Any generic Lagrangian fibration whose fiber has a common tangential direction with the cylinder at a point of the line of vertices is locally reducible to the above fibration, by a symplectomorphism preserving the cylinder.*

For instance, the projection Ω_{n+1} in theorem 2 can be reduced to this normal form.

The caustic of the Lagrangian projection of the cylinder, described above, has 2 components. One component is the image of the singular set of the Lagrangian variety. The other is the set of critical values of the projection. The theorems above imply

Corollary. *The caustic of the singularity of* Ω_{n+1} *is diffeomorphic to the bifurcation diagram of the projection* C_{n+1}.

The bifurcation diagram of the projection C_{n+1} is described in § 6.3 (see Fig. 92 and theorem 10). This diagram is a hypersurface in n-space, defined as the set of lines that are parallel to the b_n-axis in the $(n + 1)$-space of polynomials

$$x^{n+1} + b_1 x^n + \cdots + b_{n+1}; \tag{2}$$

it nontransversally intersects the variety of polynomials with a multiple root.

Example. For $n = 2$ we start in the 3-space of monic cubical polynomials with the surface of polynomials with a multiple root. It can be projected to the horizontal plane by a bundle of vertical lines (Fig. 92). The tangent plane of the cylinder is vertical at a certain point of the cuspidal edge. The bifurcation diagram of the projection C_3 is the curve in the horizontal plane consisting of the projection of the cuspidal edge of the cylinder and of the projection of the tangent plane to the cylinder (at the point where it is vertical).

Generically, the bifurcation diagram of the projection C_{n+1} is the set $\{(b_1, \ldots, b_{n-1}; b_{n+1})$: there is a b_n such that the polynomial (2) either has a root of multiplicity exceeding 2 or a root at 0 $(b_{n+1} = 0)\}$.

In order to prove the corollary, it suffices to differentiate $n-1$ times the polynomials (1), and to divide by a suitable constant. The Lagrangian variety of polynomials (1) having a root of multiplicity exceeding n can be projected onto the cylinder of polynomials (2) with a multiple root. The Lagrangian map in theorem 4 is then transformed to the map

$$(b_1, \ldots, b_{n+1}) \mapsto (b_1, \ldots, b_{n-1}, b_{n+1})$$

(i.e. 'forgetting b_n'), providing the required description of the caustic.

In order to study the Legendre projections and fronts corresponding to the Lagrangian projections above, we have to contactify the symplectic space, the Lagrangian fibration, and the Lagrangian submanifold. We choose the cotangent bundle $(p, q) \mapsto q$ as the local normal form of the Lagrangian fibration. The contactified space is then the space $\{(p, q; z)\}$ of 1-jets of functions, equipped with the contact structure $dz = p\,dq$. The Legendre variety corresponding to a given Lagrangian variety is the variety of 1-jets of the (multivalued) generating function

$$z = \int p\,dq \quad \text{(integrals along the Lagrangian variety)}.$$

The front of the Legendre variety corresponding to a family of functions (as in theorem 1) is the bifurcation diagram of zeros of the family (or, equivalently, the graph of the multivalued function on the parameters of the family whose values are the critical values of the function in the family corresponding to the given parameter values).

Theorem 5 ([8]). *Any two holomorphic vector fields that are transversal to the bifurcation diagram of zeros of any family Ξ, Ω, H (see theorems 1, 2) are locally reducible to each other, by a local holomorphic diffeomorphism of the ambient space preserving the bifurcation diagram.*

Remark. This theorem also holds in the real analytic or smooth case.

Corollary (*[8]*). *In an obstacle problem, the simple singularities of the time function are precisely the singularities of the generating functions of the Lagrangian maps A_k, D_k, E_k, Ξ_k, Ω_k, H_k, or their suspensions (suspensions are always present in the case of Ξ).*

Corollary (*[8]*). *The complements of the complex bifurcation diagrams of functions of a family Ξ_k, Ω_k, H_k are Eilenberg—MacLane spaces $K(\pi, 1)$, where π is a subgroup of finite index in the braid group on k strings.*

Example 1. The graph of a multivalued time function having singularity A_3 is shown in Fig. 66.

Example 2. An explicit parametrisation of the front Ξ_{n+1} can be obtained from the above normal form of the Lagrangian singularity, by taking into account that for the contact structure

$$dz = p\,dq - q\,dp$$

the map

$$(p, q; z) \mapsto (q, w = z + pq)$$

is a Legendre fibration. Thus, we first consider the contact space of polynomials

$$e_{2n+3} + q_1 e_{2n+1} + \cdots + q_n e_{n+2} - p_n e_{n+1} + \cdots + (-1)^n (p_1 e_2 - p_0 e_1 + z),$$

where $e_j = x^j / j!$ and

$$p_0 = -\left(q_1 p_2 + q_2 p_3 + \cdots + q_{n-1} p_n + \frac{q_n^2}{2} \right), \qquad dz = p\,dq - q\,dp.$$

The polynomials

$$\left(x - \frac{a_1}{n+3} \right)^{n+3} \cdot \frac{x^n + a_1 x^{n-1} + \cdots + a_n}{(2n+3)!},$$

having a root of multiplicity exceeding $n + 2$, form the Legendre variety Ξ_{n+1} (it is diffeomorphic to the unfurled swallowtail of dimension n and is quasihomogeneously parametrised by the coordinates $(a_1, \ldots, a_n)$ of weights $(1, \ldots, n)$).

Consider the Legendre fibration

$$(p, q; z) \mapsto (q, w = z + pq).$$

Theorem 6. *Any generic Legendre fibration of a space containing a Legendre unfurled swallowtail is locally reducible to the above one, by a local contactomorphism reducing the unfurled swallowtail to the above normal form.*

For instance, the Legendre projection generated by the Lagrangian projection in theorem 2 can be reduced to the normal form above.

The front of the Legendre projection Ξ_{n+1}, using the notation above, is the projection of the Legendre variety. Hence it can be quasihomogeneously parametrised by the functions

$$(q_1(a), \ldots, q_n(a); w(a) = z(a) + p(a)q(a)),$$

of weights $(2, \ldots, n + 1; 2n + 3)$.

This is an explicit parametrisation, but it is rather unpleasant.

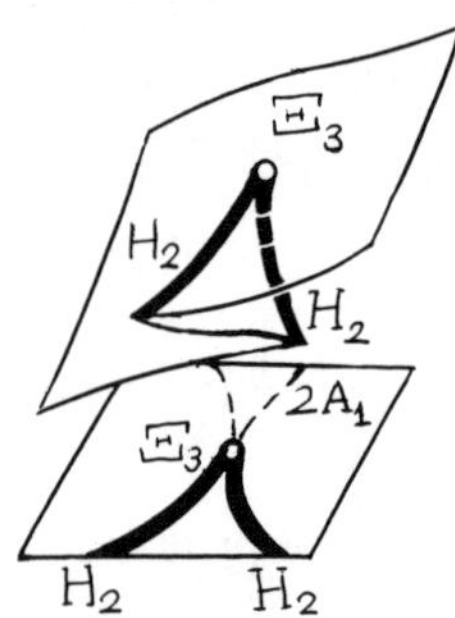

Figure 116: The caustic of the front Ξ_3

Example. By some (long) computations one can deduce, for instance, the following (cf. [23]).

Corollary. *The front Ξ_3 is diffeomorphic to the surface in (A,B,C)-space given parametrically by the equations*

$$A = s, \quad B = t(s + 7t^2), \quad C = t^3(10s^2 + 189st^2 + 945t^4).$$

In Fig. 116 this surface is shown as the graph of a multivalued function $C(A,B)$ (any vector field that is transversal to this surface at the origin can be reduced to the normal form $\partial/\partial C$, by a local diffeomorphism preserving the surface). The line of selfintersection of the front Ξ_3 is complex (it is a semicubic parabola, indicated by a dotted line in Fig. 116).

The front Ω_4 is shown in Fig. 117, which is due to O.P. Shcherbak [144].

Warning. In Fig. 117 the sections are defined by the equations $q_1 = \text{const}$. However, for $n > 1$ a generic function in a neighborhood of the vertex of the front Ω_{n+1} cannot be reduced to q_1 by a diffeomorphism preserving the front. The problem of computing the normal form to which a generic function in a neighborhood of a front Ω can be reduced is still unsolved, as far as I know.

The explicit parametrisation of the front Ω_{n+1} uses the Legendre fibration

$$(q_1, \ldots q_n; p_1, \ldots, p_n; z) \mapsto (q_1, \ldots, q_{n-1}; p_n; w),$$

where $w = z + p_1 q_1 + \cdots + p_{n-1} q_{n-1} - p_n q_n$; the contact structure being $dz = p\,dq - q\,dp$.

We start with the space of polynomials

$$e_{2n+1} + q_1 e_{2n} + \cdots + q_n e_{n+1} - p_n e_n + \cdots + (-1)^n(p_1 e_1 - z),$$

where $e_j = x^j/j!$, equipped with the usual contact structure $dz = p\,dq - q\,dp$. The polynomials having a root of multiplicity exceeding $n + 1$ form the Legendre variety

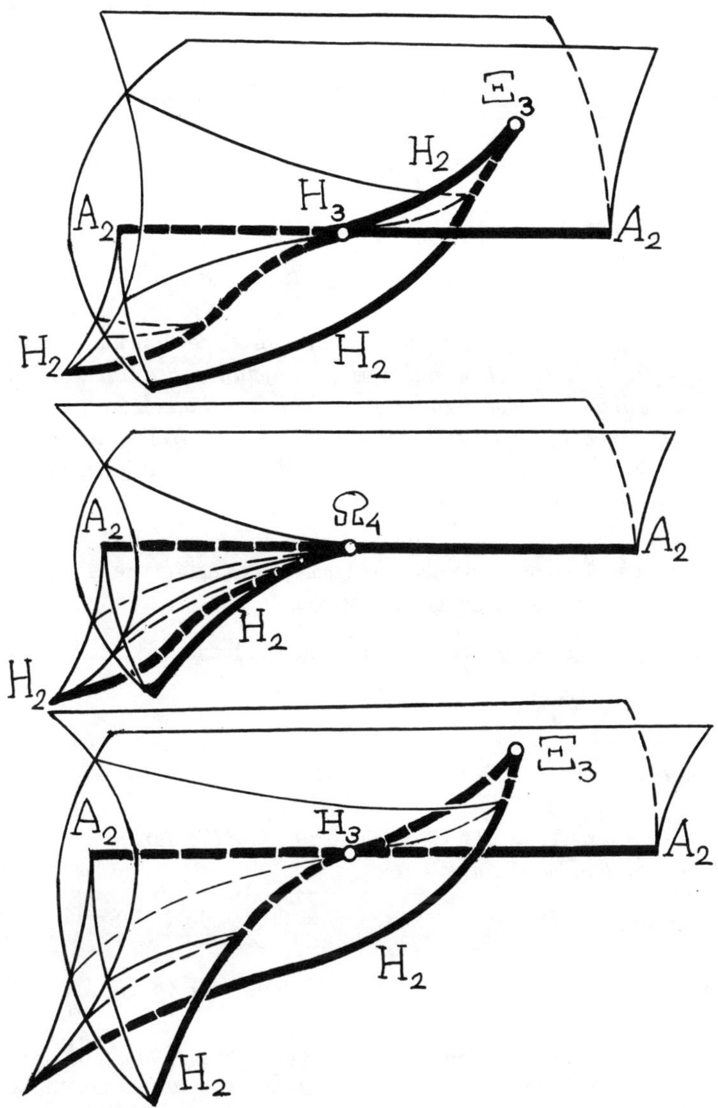

Figure 117: The front Ω_4 as a perestroika of its sections

Ω_{n+1}. It is diffeomorphic to the product of the unfurled swallowtail and the line. Writing the polynomials as

$$(x + b_1)^{n+2} \frac{x^{n-1} + a_1 x^{n-2} + \cdots + a_{n-1}}{(2n+1)!},$$

we obtain a quasihomogeneous parametrisation of the Legendre variety (the weights of a_j, b_j are equal to j).

Consider the Legendre fibration

$$(p, q; z) \mapsto (q_1, \ldots, q_{n-1}; p_n; w = z + p_1 q_1 + \cdots + p_{n-1} q_{n-1} - p_n q_n).$$

Theorem 7. *Any generic Legendre fibration of a space containing a Legendre variety diffeomorphic to the product of an unfurled swallowtail and a line is locally reducible to the above fibration, by a local contactomorphism reducing the product to normal form, as described above, in a neighborhood of a point of the line of vertices of the product at which the tangent plane of the Legendre variety is vertical (has a line in common with the tangent plane of the fiber).*

The front of the Legendre projection Ω_{n+1} can, using the notation above, be parametrised by the quasihomogeneous functions

$$(q_1(a, b), \ldots, q_{n-1}(a, b); p_n(a, b); w(a, b))$$

of weights $(1, \ldots, n-1; n+1; 2n+1)$.

Example. By some (long) computations one can, for instance, deduce the following.

Corollary. *The front $\Omega_3 = H_3$ is diffeomorphic to the surface in (A, B, C)-space given by the parametric equations*

$$A = s, \quad B = 10t^3 - 3st^2, \quad C = 60t^5 - 45st^4 + 8s^2 t^3.$$

The front Ω_3 is shown in Fig. 118. It is diffeomorphic to the discriminant of the icosahedron symmetry group (Fig. 102).

A vector field that is transversal to the tangent plane of this surface at the origin $(\dot{C} \neq 0)$ is locally reducible to the normal form $\partial/\partial C$, by a diffeomorphism preserving the surface. Hence the singularity of the multivalued function $C(A, B)$, whose graph is our function, does not depend on the choice of the vector field.

The caustic consist of 2 smooth curves, which are tangent of order 3 (it is the bifurcation diagram of the projection C_3, see Fig. 92).

A generic function in a 3-space containing an H_3 front can be reduced to the form $A + \text{const}$, by a local diffeomorphism preserving the surface (this follows from the usual theory of convolution of invariants of the reflection group).

Hence the decomposition of the graph of the multivalued time function into momentary fronts (level sets of the time function) is reduced to normal form in a neighborhood of a singularity H_3. Our 3-space may be thought of as being space–time

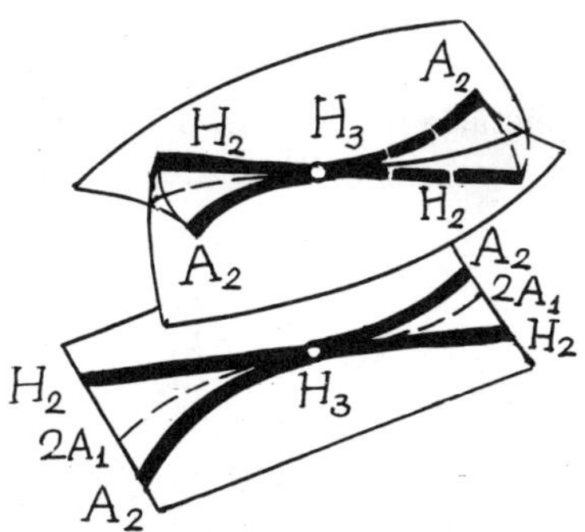

Figure 118: The front H_3 and its caustic

(time$=A$), and the sweeping of the surface of the front by the curves $A=$ const
is diffeomorphic to the H_3 perestroika of the momentary fronts moving in the plane
containing the obstacle.

The relation of the front Ω_3 to the icosahedron symmetry group allows us to write
normal forms of Ω_3 perestroikas in higher dimensions. Add new coordinates $D, E, \ldots$
and extend the discriminant cylindrically. The normal form of the 'time' function
then is

$$T = D \qquad \text{or} \qquad T = A \pm D^2 \pm E^2 \pm \cdots$$

(a generic function in a space containing the cylinder can be reduced to this local normal form by a diffeomorphism preserving the cylinder and by addition of a constant).

Unfortunately, there is no reflection group corresponding to Ξ_n, $n > 2$, or Ω_n,
$n > 3$. Hence we do not know local normal forms for functions in a space containing
the corresponding fronts. In these cases we do know normal forms of multivalued
time functions, but not those of momentary fronts and their perestroikas. The first
unsettled case is Ξ_3.

The theory above describes most of the singularities in an obstacle problem, but
not all of them (even in the case of an obstacle bounded by a smooth surface in
euclidean 3-space), since we have neglected the singularities of the family of geodesics
on the surface of an obstacle, and since we have assumed that the distance to the
'source' manifold be smooth.

The most complicated singularity (H_4) occurs at a point of a straight line tangent
to the surface of the obstacle at a parabolic point and having an asymptotic direction.
(In a generic obstacle problem the direction of a geodesic in a bundle of extremals on
the surface of the obstacle and an asymptotic direction coincide at certain isolated
parabolic points.)

The graph of the multivalued time function is diffeomorphic to the discriminant
of H_4 in a neighborhood of the 'focal' point of the straight line issuing from the
above parabolic point along an asymptotic direction—this is the final conclusion of
Shcherbak's theory [144].

The singularity Ω_4 occurs at a point of the biasymptotic line on the surface of the obstacle that is tangent to a geodesic in our bundle of geodesics on the surface of the obstacle. (For the generic obstacle problem in 3-space the tangents are biasymptotic at certain isolated points of the surface; these depend on the initial condition.)

Chapter 8

Transformation of waves defined by hyperbolic variational principles

In this chapter we consider the geometrical optics of waves defined by linear differential equations or systems. In the linear theory, waves of different kinds (say, longitudinal and oblique waves) usually propagate independently. However, in nonhomogeneous media a transformation (or 'conversion') of waves of different kinds is typical at certain interior points of the domain. This 'interior scattering' of waves also occurs in homogeneous media (Hamilton's conical refraction in crystals). However, the geometry of the interior scattering in generic nonhomogeneous media is rather different from the geometry of Hamilton's conical refraction, as will be seen below[1].

It is well known that the global qualitative properties of Hamiltonian differential equations are rather different from those of ordinary differential equations defined by generic vector fields (for instance, attractors, which are so important in the latter case, are absent in the former case). Still, locally a Hamiltonian vector field is as simple as a generic vector field: both can be reduced to the same trivial normal form in a neighborhood of any nonsingular point, by a diffeomorphism.

There should be a similar difference between generic partial differential equations and those governed by variational principles (mysteriously, the latter class includes most systems of partial differential equations of physical interest).

In the theory of hyperbolic partial differential equations mathematicians have never distinguished between variational and generic systems. Indeed, symmetrisation reduces a generic system to the same form as a variational system, in a point of strict hyperbolicity. A generic system is strictly hyperbolic at generic points, hence at such

---

[1] Technically speaking, *homogeneous* media correspond to *nongeneric* contact structures and generic *nonhomogeneous* media correspond to *generic* contact structures of the spaces, containing the *same* quadratic conical singularity on the light hypersurface.

points there is no essential distinction between variational and generic systems.

However, in the global consideration, points of nonstrict hyperbolicity become unavoidable (just like singular points of a vector field are unavoidable in the global, rather than the local, consideration of the field).

The global difference between the behavior of generic and Hamiltonian dynamical systems manifests itself locally at the singular points. Similarly, in the theory of hyperbolic partial differential equations the behaviors of rays and wave fronts in generic systems and in variational systems are rather different in neighborhoods of the 'singular' points of nonstrict hyperbolicity, while at the remaining points the propagation of waves is similar in both cases.

The main results in this chapter are presented in §§ 8.3–8.4, in which we describe the normal forms of singularities of the light hypersurface, of ray systems, and of wave fronts. For simplicity, only the case of Petrovsky linear hyperbolic systems of partial differential equations, derived from variational principles, are discussed.

The reader will be able to extend most of the results to the interior scattering of short waves defined by variational principles containing small parameters, like the Schrödinger equation.

Mathematically, the main results in this chapter describe the normal forms of singularities of hypersurfaces in contact manifolds. Since such objects are also encountered in other domains of mathematics and physics, these results have applications outside the theory of propagation of waves (for instance, in the theory of implicit ordinary differential equations and in the theory of bifurcation of relaxation systems with 1 fast and 2 slow variables).

8.1   Hyperbolic systems and their light hypersurfaces

The geometrical optics of rays and fronts defined by a system of hyperbolic partial differential equations is the geometry of a hypersurface in a contact space of the projectivised cotangent bundle of space–time. This hypersurface, called the light hypersurface, is the set of zeros of the (principal) symbol. In the theory of partial differential equations, the characteristics of this hypersurface in the contact manifold of contact elements are, strangely, called the 'bicharacteristics'.

These curves (and their projections to space–time and physical space) are called the physical rays. The Legendre submanifolds of the light hypersurface, projected to space–time, define in space–time the hypersurface of the 'big front'. This hypersurface describes the propagation of waves: its sections by the isochrones are the momentary fronts.

To fix notations, I recall the well-known definitions of light hypersurface and of hyperbolic system of partial differential equations.

We start with a system of linear homogeneous partial differential equations with constant coefficients:

$$P(D)u(q) = 0.$$

We fix 3 integers:

$m = \dim u$, the number of equations (and unknown functions);

$d = \operatorname{ord} P$, the order of the differential operator P;

$n = \dim q$, the number of arguments.

In coordinates, P is a square m-matrix with as entries homogeneous polynomials of degree d in $(\partial/\partial q_1, \ldots, \partial/\partial q_n)$. In other words, the argument q is a point in the *base manifold* $\mathbf{R}^n$, the vector $u(q)$ belongs to the *fiber vector space* $\mathbf{R}^m$, the unknown vector function u is a section of the (trivial) vector bundle (and P sends it into another trivial vector bundle, of the same dimension).

Since the operator P is translation invariant, we consider the eigenfunctions of the translations, which are the harmonic waves $u = e^{i(p,q)}w$, where $p \in \mathbf{R}^{n*}$ is called the *wave vector* and $w \in \mathbf{R}^m$ is called the *amplitude vector*. Applying P to a harmonic wave gives

$$P(D)u = \sigma(ip)u,$$

where σ is a square m-matrix whose entries are homogeneous polynomials of degree d in $(p_1, \ldots, p_n)$. This matrix polynomial is called the *matrix symbol* of the system. In terms of σ the operator of our system takes the form

$$P(D) = \sigma\left(\frac{D}{i}\right) = \sigma\left(\frac{1}{i}\frac{\partial}{\partial q_1}, \ldots, \frac{1}{i}\frac{\partial}{\partial q_n}\right).$$

The matrix symbol is a homogeneous polynomial map from the space of wave vectors to the space of linear operators on the fiber.

Definition 1. The *symbol* of the system $P(D)u = 0$ is the determinant of the matrix symbol,

$$\sigma = \det \sigma.$$

Thus, the symbol is a homogeneous polynomial of degree md, defined on the space of wave vectors $\mathbf{R}^{n*}$.

Definition 2. The *Fresnel cone* is the set of zeros of the symbol. It is an algebraic cone in the space of wave vectors, and is intrinsically associated to the system.

Remark. The equation $\sigma = 0$ is also called the *dispersion relation* or the *characteristic equation*. The Fresnel cone is also called the *characteristic cone*.

Definition 3. The *Fresnel hypersurface* is the set of zeros of the symbol in the projective space

$$\mathbf{R}P^{n-1} = (\mathbf{R}^{n*} \setminus 0)/(\mathbf{R} \setminus 0)$$

of contact elements. It is a projective algebraic hypersurface of degree md.

Remark. It is sometimes useful to consider the sphere of (co)oriented contact elements and the Fresnel hypersurface of oriented contact elements living in this sphere.

Definition 4. An algebraic hypersurface of degree N is called *hyperbolic at a point* if all its intersections with real straight lines through this point are real.

The hypersurface is called *strictly hyperbolic at a point* if all these intersection points are distinct.

The point we start with in the definition of hyperbolicity is called a *time-like point*.

A hyperbolic hypersurface is generically hyperbolic at many time-like points.

In the linear space corresponding to the projective space, a hyperbolic hypersurface is represented by a cone, which is also called hyperbolic. The time-like points are represented by straight lines, called *time-like directions*.

Definition 5. A system of partial differential equations is called *hyperbolic* (with respect to a time-like direction) if its Fresnel hypersurface is hyperbolic (at the corresponding time-like point).

Remark. A time-like point is a contact element in the base space. Hence a system is hyperbolic with respect to a (*time*) *function* in the base space, rather than with respect to a line.

We will now consider a system with variable coefficients. Written in the local coordinates $(q_1, \ldots, q_n)$ in the base manifold, the entries of the matrix $P = P(D, q)$ are now nonhomogeneous polynomials of degree d in $(\partial/\partial q_1, \ldots, \partial/\partial q_n)$, with coefficients that are smooth functions of q. The matrix depends also on the (local) trivialisations of the vector bundles introducing the coordinates $(u_1, \ldots, u_m)$, $(P_1, \ldots, P_m)$ in the fibers.

Definition 6. The *principal (matrix) symbol* of a linear system of partial differential equations with variable coefficients is the (matrix) symbol of the system with constant coefficients obtained by fixing a point in the base manifold and neglecting in the operator the terms containing lower order derivatives.

Remark. Thus, the principal (matrix) symbol is a homogeneous polynomial function (operator) of degree md (d), defined on the cotangent space $T_q^* B$ of the base manifold (the operator acts on the corresponding fiber). This function (operator) does not depend on either the choice of local coordinates in the base manifold, nor on the trivialisations of the fiber bundles. Hence we obtain the *principal symbol map*

$$\sigma : T^* B \to \mathrm{Hom}$$

and the *principal symbol function*

$$\sigma : T^*B \to \mathbf{R}, \qquad \sigma(p,q) = \det \boldsymbol{\sigma}(p,q),$$

which are homogeneous polynomials in the moments p (the values of σ belong to the determinant line bundle over B).

Definition 7. The *light cone* (of a system with variable coefficients) is the union of the Fresnel cones of the principal parts of the frozen systems at all points of the base manifold; it is defined by the equation $\sigma(p,q) = 0$. The same equation defines the *light hypersurface* in the space of the projectivised cotangent bundle of the base manifold; it is the union of the Fresnel hypersurfaces of the principal parts of the frozen system.

This light hypersurface

$$H \subset PT^*B$$

consists of the directions of the wave fronts at the different points of the base manifold which correspond to the harmonic wave solutions of the principal parts of the frozen system.

Definition 8. A system of linear partial differential equations with variable coefficients is called *hyperbolic with respect to a time function* defined on the base manifold if its light hypersurface is hyperbolic with respect to the differential of this function at every point of the base manifold.

Remark. One can also define hyperbolicity with respect to a foliation of codimension 1, or even with respect to a distribution of contact elements on the base manifold (integrable or not).

The light hypersurface is the union of algebraic projective hypersurfaces belonging to different fibers of the bundle of contact elements of the base space. Generically, these algebraic hypersurfaces are nonsingular (in this case they are strictly hyperbolic). But at certain points of the base manifold they may acquire singularities. The study of these singularities for generic variational hyperbolic systems is the main goal of this chapter.

Example. Consider a field $u(t,x)$ with values in the euclidean space $\mathbf{R}^m$, defined in the space–time $\{(t,x)\} = \mathbf{R}^n$, $n = 1 + D$, by a variational principle with Lagrangian $L = T - U$, where $T = \int (\partial u / \partial t)^2 / 2 \, dx$ is the kinetic energy and where the potential energy is an integral of a positive definite quadratic form in the first derivatives of the field with respect to the (spatial) x coordinates. The coefficients of the form are, in general, functions of t and x, but we first consider the case of constant coefficients.

The Euler—Lagrange equation takes the form

$$Pu = 0, \qquad P = \frac{\partial^2}{\partial t^2} + A,$$

where A is a symmetric matrix whose entries are second order differential operators with respect to the x variables, $A = a(\partial/\partial x_1, \ldots, \partial/\partial x_D)$.

Lemma. *The Euler—Lagrange system is hyperbolic if the potential energy is positive definite.*

Proof. Denote by $q = (t, x)$ the points of space–time $\mathbf{R}^n$ and by $p = (\omega, k) \in \mathbf{R}^{n*}$ the corresponding wave vectors. The symbol of the system is

$$\sigma(p) = \det(a(ik) - \omega^2 E).$$

The matrix $a(ik)$ is positive definite for any $k \neq 0$ (since the potential energy is positive definite).

Regard the time function as a vector $p_0 = (1, 0) \in \mathbf{R}^{n*}$. Hyperbolicity (with respect to the time function) means that all the $2m$ roots λ of the characteristic equation $\sigma(p + \lambda p_0) = 0$ are real (for any p). But the roots of the characteristic equation

$$\det(a(ik) - (\omega + \lambda)^2 E) = 0$$

are $\lambda = -\omega \pm \sqrt{\alpha_j(k)}$, where $\alpha_j(\mathrm{k})$ are the eigenvalues of the matrix $a(ik)$. Since this matrix is positive definite (or zero), all $2m$ roots λ are real.

Essentially the same proof works in the case of a variational principle with variable coefficients: to write down the symbol it suffices to start with the variational principle with frozen coefficients.

A variational principle defines a map from the space of the cotangent bundle of the base manifold (which is the space–time in the preceding example) to the space of *symmetric* $(m \times m)$-matrices (more precisely, to the symmetric tensor square of the dual space of the fiber). This map is homogeneous (the matrix entries are homogeneous polynomials of degree $d = 2r$, if the variational principle involves derivatives of order r), Conversely, any symmetric matrix with these properties is the principal matrix symbol of an Euler—Lagrange system for a variational principle with quadratic Lagrangian.

Definition. A variational principle is *hyperbolic* if its Euler—Lagrange system is hyperbolic (the Fresnel surfaces are hyperbolic) with respect to some 'time' function.

8.2 Singularities of light hypersurfaces of variational systems

The Fresnel hypersurfaces of a generic hyperbolic variational principle are singular at certain points of the base manifold (at which the hyperbolicity is nonstrict). Now we will study the singularities of the Fresnel and light hypersurfaces occurring in generic hyperbolic variational principles.

Fix a 'signature' (m, d, n): the number of unknowns is m, the order of the differential operator is d, and the number of independent variables is n (in the global formulation, a vector bundle is fixed in addition). The quadratic variational principles with a fixed signature form a linear function space. The hyperbolic variational principles form a domain with boundary in this space. We will study the interior points of this domain.

The interior points of the space of hyperbolic polynomials are the strictly hyperbolic polynomials. The situation is different for hyperbolic systems, and, in particular, for the hyperbolic Euler—Lagrange systems defined by variational principles (see [182]–[186]).

Example. The quadratic Lagrangian of § 8.1, which is the difference between the positive definite 'kinetic energy' and the positive definite 'potential energy', is an interior point of the domain of hyperbolic variational principles of signature $(m, 2, n)$. If $m > 2$, it can define a nonstrictly hyperbolic system, since the positive definite symmetric matrix a can have multiple eigenvalues.

In order to avoid unnecessary complications, we will now consider the local situation in which the coordinates in the fibers are fixed (the bundle is trivialised). The principal matrix symbol is then a homogeneous map

$$\sigma : T^*B \to S^2 \mathbf{R}^m \approx \mathbf{R}^{m(m+1)/2},$$

where $S^2 \mathbf{R}^m$ is the space of symmetric m-matrices, and where the components of σ are homogeneous polynomials of degree d along each cotangent space.

Definition. A polynomial map of degree d from the real line to the space of quadratic forms in m variables,

$$f : \mathbf{R} \to F = S^2 \mathbf{R}^m \approx \mathbf{R}^{m(m+1)/2},$$

is called a *hyperbolic map* if the equation $\det f(s) = 0$ has md real roots (counted with multiplicities).

All such maps form a (semi-algebraic) space,

$$\mathrm{Hyp}\,(m, d).$$

A variational hyperbolic system over the space–time B of dimension $n = D + 1$ defines (at any point of B) a map of the sphere:

$$F : S^{D-1} \to \mathrm{Hyp}\,(m, d).$$

The sphere belongs to a D-dimensional hyperplane in T^*B that is at the origin transversal to the vector $v = dt$ (where t is the time function).

In order to define the value of F at a point p of the unit sphere of the hyperplane, we regard the principal symbol at the point $p + sv$ as a polynomial in s,

$$(F(p))(s) = \det \boldsymbol{\sigma}(p + sv).$$

Hence the following problems arise:

1. *compute the homotopy groups*

$$\pi_i(\mathrm{Hyp}\,(m, d));$$

2. *find out which elements of these groups are representable by hyperbolic variational (pseudo)differential systems;*

3. *count the connected components of the space of hyperbolic variational systems with a fixed signature* (m, d, n).

The similar problems for bundles over B are interesting in case the above three problems have nontrivial answers. One can conjecture that the presence of nonstrict hyperbolic points is necessary for the realisation of certain classes, and that one can use the singularities of the light hypersurface in order to define the characteristic classes of hyperbolic variational principles.

The light hypersurface is the pre-image of the variety of degenerate matrices in PT^*B under the map given by the principal matrix symbol.

In order to study the singularities of the light hypersurface, we will first describe the singularities of the corresponding universal object: the variety of degenerate symmetric matrices.

This hypersurface is a singular cone, stratified by rank and signature. Let $N \approx \mathbf{R}P^{m(m+1)/2-1}$ be the projective space of nonzero quadratic forms in m variables, considered up to nonzero scalar multiples. The following result is well known:

Theorem. *The projective space N admits an algebraic filtration $N \supset N_1 \supset N_2 \supset \cdots$, where N_r is the variety of classes of forms of corank at least r. The codimension of N_r in N is equal to $r(r + 1)/2$. The trace of N_1 on any manifold of dimension $r(r + 1)/2$ and transversal to the smooth manifold $N_r \setminus N_{r+1}$ in N is, at a point of N_r, locally diffeomorphic to the variety of degenerate quadratic forms in the linear space of quadratic forms in r variables.*

Example. The codimension of N_2 is 3. The trace of the variety of degenerate forms on a 3-manifold that is transversal to the variety of forms of corank 2 is locally diffeomorphic to the quadratic cone $b^2 = ac$ (Fig. 119).

Indeed, this is the equation of the cone of degenerate quadratic forms $ax^2 + 2bxy + cy^2$.

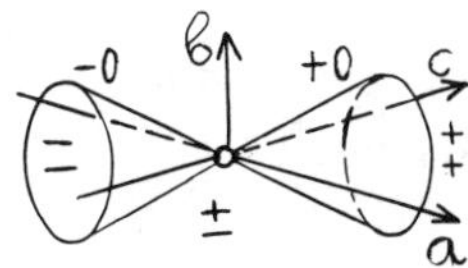

Figure 119: The stratification of the space of binary quadratic forms

Consider now the smooth maps $f : X \to N$ from a manifold X of dimension k into the manifold of classes of quadratic forms. A typical map f is transversal to the above rank filtration. Hence the theorem above implies:

Corollary. *There exists an open dense set of 'typical' maps f that are transversal to all manifolds $N_r \setminus N_{r+1}$, such that:*

1. *the forms of corank r occur in the image of f only for $r(r+1)/2 \le k$;*

2. *the pre-image $f^{-1}(N_1)$ of the set of degenerate forms is a hypersurface in X whose singular points form a variety of codimension 2 in the hypersurface;*

3. *the above hypersurface is at its generic singular points locally diffeomorphic to the product of a 2-dimensional quadratic cone and a smooth manifold, while the nongeneric points form a variety of codimension 5 in the hypersurface.*

Indeed, the generic singular points belong to $f^{-1}(N_2 \setminus N_3)$, and the nongeneric points belong to $f^{-1}(N_3)$, which has codimension 6 in X.

The light hypersurface is the pre-image of the variety N_1 of degenerate forms under the (projectivised) map given by the principal matrix symbol. Hence the corollary provides information about the singularities of light hypersurfaces defined by generic variational principles.

Remark. The map defined by the principal matrix symbol of a differential equation cannot be arbitrary, since the principal matrix symbol is a homogeneous polynomial with respect to the cotangent vector. However, the preceding genericity arguments work also in this case, since there exist deformations of the variational principle such that the image of the derivative of the principal matrix symbol with respect to the parameters covers the whole tangent space to N (for any point in the cotangent bundle outside the zero section).

Example. Consider the hyperbolic variational principles in the space–time of dimension $n = D + 1 = 2$ (the physical space dimension being $D = 1$).

The above theory implies:

Theorem. *The typical singularities of light surfaces in hyperbolic variational principles with 1 physical space dimension are diffeomorphic to the quadratic cones (with local normal form $x^2 + y^2 = z^2$, where (x, y, z) are local coordinates in the manifold of contact elements of the space–time plane).*

Hence the singularities are isolated points, and the corresponding 'interior scattering' occurs at special time moments and at certain singular points.

The same normal form $x^2 + y^2 = z^2$ describes (up to diffeomorphisms) the simplest singularity of the light hypersurface in the manifold of contact elements of a higher-dimensional space–time (with local coordinates $x, y, z, u, v, \ldots$).

For $D = 2$ (two space variables) these singular points form a 2-dimensional 'manifold of vertices' on the 4-dimensional light hypersurface in the 5-dimensional space of contact elements. In this dimension there are no other singularities. The projection of the manifold of vertices to 3-dimensional space–time is a 2-dimensional variety. Hence scattering is possible, at any moment, at the points of a time-dependent curve in physical 2-space (equivalently, at any point of the 2-space and at specific time moments, depending on the point).

For $D = 3$ (three space variables) the dimension of the light hypersurface is 6, and hence the dimension of the 'variety of vertices' of a generic variational principle is 4. Besides these simple singularities of quadratic cone type along the variety of vertices, there can exist a curve of singularities of type N_3. The projection of the variety of vertices to space–time can cover the whole 4-dimensional space–time. Hence scattering is possible at any point and at any time moment (however, of course the direction of the singular wave is special, as it was in the spaces of smaller dimension).

Singularity of the light hypersurface indicates that something unusual is happening with the propagation of the waves. In order to understand the singularities of the corresponding ray systems and fronts, we need information on the singularity of the light hypersurface as a surface in contact space. Hence we are led to the following problem: reduce to local normal form a singular surface that is diffeomorphic to a quadratic cone in a contact space, by a contactomorphism.

8.3   Contact normal forms of singularities of quadratic cones

Consider a hypersurface that is diffeomorphic to the quadratic cone $x^2 + y^2 = z^2$ in a contact space of dimension $2n + 1$.

Theorem. *The pair formed by a cone and a generic contact structure can be reduced (in a neighborhood of the vertex of the cone) to the normal form*

$$p_1^2 \pm q_1^2 = z^2 + cz^3 \quad (n = 1), \qquad p_1^2 \pm q_1^2 = q_2^2 \quad (n > 1),$$

and the contact structure takes the Darboux normal form

$$dz + \frac{p\,dq - q\,dp}{2} = 0.$$

Remark. This theorem was proved at the level of formal series in [182] for $n = 1$ and in [183] for $n > 1$. In the analytic (or holomorphic) case the series defining the reducing diffeomorphism generically diverge.

There probably exist C^∞-diffeomorphisms reducing the conical singularity to the above normal form (an outline of a proof for $n = 1$ can be found in [182]; details have not been published, however).

There is an interesting relation between reduction to normal form of contact structures in a 3-dimensional neighborhood of the vertex of the quadratic cone and the theory of normal forms of equivariant planar vector fields at the singular points.

First, it suffices to reduce to normal form the restriction of the contact structure to the surface of the cone (for details see [182] or [8]). In order to reduce to normal form a 1-form on the surface of a cone, consider the twofold 'covering' of the cone $x^2 + y^2 = z^2$ ($z \geq 0$) by the (u, v)-plane:

$$x = u^2 - v^2, \quad y = 2uv, \quad z = u^2 + v^2.$$

Lemma 1. *A differential 1-form in 3-space induces on the plane a form $P\,du + Q\,dv$ whose coefficients are odd functions and whose differential vanishes at the origin. Conversely, any planar form with these properties is induced by a 1-form in space.*

Let us call a diffeomorphism of the (u, v)-plane *odd* if it commutes with the change of sign of both variables simultaneously: $(u, v) \mapsto (-u, -v)$.

Lemma 2. *Any odd (local) diffeomorphism of the (u, v)-plane is induced from a diffeomorphism of the ambient 3-space preserving the cone.*

These lemmas reduce our problem to the local orbital classification of the odd vector fields having trace (sum of eigenvalues at the origin) 0 with respect to the group of odd diffeomorphisms.

The standard theory of normal forms of vector fields (see, for instance, [17]) implies

Lemma 3. *A normal (or C^∞) odd diffeomorphism and a multiplication by an even function reduce a generic odd vector field with trace 0 either to the hyperbolic normal form*

$$\begin{cases} \dot{u} = u(1 \quad +uv + c(uv)^2), \\ \dot{v} = v(-1 \quad +uv + c(uv)^2), \end{cases}$$

or to the elliptic normal form

$$\begin{cases} \dot{u} = v \quad +u(u^2+v^2)+cu(u^2+v^2)^2, \\ \dot{v} = -u \quad +v(u^2+v^2)+cv(u^2+v^2)^2. \end{cases}$$

Returning from vector fields to 1-forms on the cone, and then to 1-forms in 3-space, one can obtain from these formulas the formulas in the theorem. This reduction also explains why the modulus c cannot be left out in the theorem: it is intrinsically related to the singularity of the corresponding planar odd vector field.

Remark. The reduction above is also useful in order to understand the behavior of the characteristics on the surface of the cone: they correspond to the trajectories of the vector fields above.

The elliptic case corresponds to the '+' sign in the theorem. In a linear approximation, the trajectories in the (u,v)-plane are closed circles. The characteristics on the cone are spirals, close to circular 'horizontal' sections $z =$ const. The height z is monotone along the characteristics (on one halfcone the characteristics slowly approach the vertex, on the other they slowly move away from it).

The hyperbolic case corresponds to the '−' sign in the theorem. In this case the characteristics on the surface of the cone are very similar to vertical sections $z =$ const. Two of them are smooth curves (analytic in the analytic case) containing the vertex. These 'separatrices' on the surface of the cone correspond to the coordinate axes $u = 0$ and $v = 0$ in the plane. Their existence follows from the Hadamard—Perron theorem (see, for instance, [17]).

In the above corollaries of the normal forms theorem convergence of the series defining the reducing diffeomorphisms is not needed, since only reduction of the first few terms in the Taylor series is used. Thus, in order to obtain topological information on the behavior of characteristics, it is not necessary to prove the complete normal forms theorem: it suffices to normalise the first few terms in the Taylor series. This normalisation requires only a finite amount of computations, which I will omit here (see [182]).

Remark. The normal forms of a cone in a contact 3-space can be interpreted as describing a generic singularity of an implicit differential equation depending on a parameter.

Generically, the surface $F(x,y,p) = 0$ defined by a generic implicit equation in the contact space of 1-jets of functions of one variable (equipped with the natural contact structure $dy = p\,dx$) is nonsingular.

But if the equation depends on a parameter, for certain parameter values the surface acquires a singularity. In generic 1-parameter families only the simplest singularities occur, namely Morse conical points.

Hence, in the context of the theory of implicit ordinary differential equations a conical singularity occurs at the moment of perestroika only. The natural problem thus becomes that of studying the bifurcation: we have to study also the equations corresponding to two- and one-sheet hyperboloids close to the surface of a cone. This can also be done by the methods of [182] (see there § 2, remark 3).

In the theory of relaxation oscillations our problem occurs at the moment of perestroika in generic families of systems with one fast and two slow variables and depending on a parameter. The contact structure in the 3-dimensional phase space is the field of planes generated by the (vertical) direction of the fast motion and an (arbitrary) direction of the small perturbation field. The fast relaxation sends a phase point along a vertical line to the slow surface on which the motion of fast velocity vanishes. The slow evolution along this surface follows the characteristics of the slow surface in our contact space.

Generically, the slow surface is smooth. However, if the system depends on a parameter, then for certain parameter values the slow surface acquires a Morse singularity (a quadratic cone). As in the case of implicit ordinary differential equations, the natural problem is to study the complete perestroika. The above normal form of the cone provides the core of the solution to the perestroika problem.

8.4 Singularities of ray systems and wave fronts at nonstrict hyperbolic points

Here we will apply the above normal form of the singularities of light hypersurfaces to the study of propagation of waves defined by hyperbolic variational principles.

We will first consider the case $D = 1$, of one physical space variable. The light surface in the 3-space of contact elements of 2-dimensional space–time generically has quadratic cone singularities, which are reducible to the normal forms given in § 8.3 by contactomorphisms.

Theorem. *The elliptic case (the '+' sign in the normal form of § 8.3) is impossible for hyperbolic systems.*

Proof. The light hypersurface lives in the space of the projectivised cotangent bundle $PT^*\mathbf{R}^2 \to \mathbf{R}^2$ over the space–time plane. The fibers of this bundle are Legendre submanifolds. The tangent line to the fiber containing the vertex of the cone belongs to the contact plane at the vertex. In the elliptic case this contact plane $(dz = 0)$ intersects the real cone at its vertex only. Hence some of the nearby fibers do not intersect the light surface in a neighborhood of the singular point, while the others intersect it twice. Thus, the characteristic equation must have complex roots at certain points of the space–time plane, and the system cannot be hyperbolic.

Now we will consider the hyperbolic case (the '−' sign in the normal form of § 8.3).

Theorem. *The projection of the union of the separatrices/characteristics on the light surface of the cone to the space–time plane (along the fibers of the cotangent fibration)*

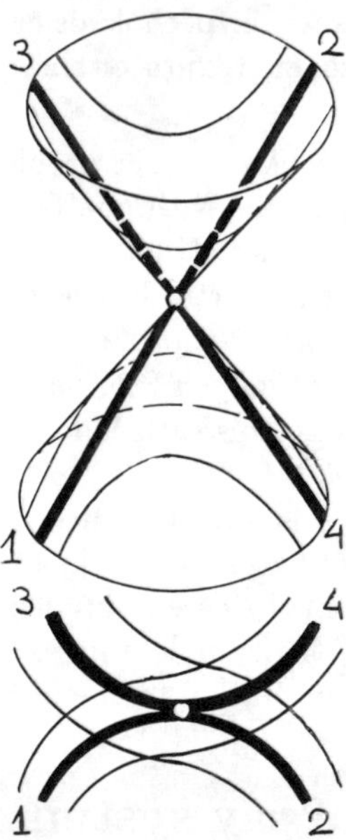

Figure 120: The characteristics of the light surface, and their projections to space–time

is formed by 2 smooth (analytic) quadratically tangent curves (1–2) and (3–4). The projections of the characteristics of the first (second) family are the curves (1–4) (respectively, (2–3)) (Fig. 120).

This follows from the fact that the tangent lines to the separatrices at the vertex belong to the contact plane, and hence are coplanar with the vertical direction (tangent to the fiber).

The quadratic deviation of one projection from the other is a manifestation of the nondegeneracy of the contact structure.

We will call the projections to space–time of the (bi)characteristics the *rays* (in this dimension they are also the big fronts). The two families of rays corresponding to the halfcones touch each other at the singular point (at the projection of the vertex of the cone), and are otherwise smooth (analytic).

Consider the 1-parameter family of rays containing the singular ray through the singular point. The singular ray is polygonal: it consists of two smooth (analytic) parts having a common tangent at the singular point but distinct curvatures. The nearby rays are smooth, but, of course, the presence of a singularity manifests itself in a scattering behavior of the family at the singular point.

Consider a smooth (analytic) line transversal at some point (at finite distance from the singular point) to the 'incoming' part of the singular ray of the first family.

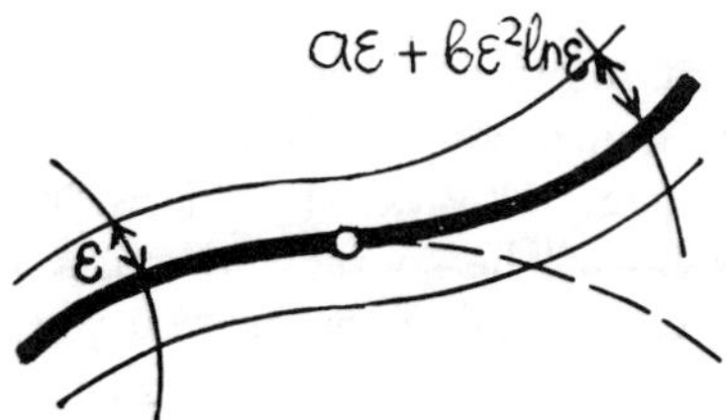

Figure 121: The scattering of rays in the space–time plane

Consider a similar line intersecting the 'outgoing' part of the singular ray (Fig. 121).

The rays of the first family transversally intersect both lines, and hence define a continuous 'scattering map', sending the point of intersection of the (incoming) ray and the first line to the point of intersection of the same (outgoing) ray and the second line. Our family of rays looks like a family of parallel lines. In a neighborhood of a point of an incoming singular ray it can be transformed into such a family by a diffeomorphism (in essence, this is done by the first transversal line). The same family of rays can also be transformed into a family of parallel lines in a neighborhood of a point of the outgoing ray. These two diffeomorphism do not, however, agree. The *scattering map* measures the extent of disagreement of the two parametrisations of the same set of rays.

Let w be the distance between the point of the incoming ray to the singular ray, measured along the transversal line of incoming points. If the family were smooth, the distance w' between the outgoing point of the same ray and the singular ray, measured along the transversal line of outgoing points, would be a smooth function of w, and its Taylor series would be $w' = aw + bw^2 + \cdots$. The normal forms of § 8.3 imply a singularity in the asymptotic development, namely $w' = aw + bw^2 \ln w + \cdots$. The logarithm is a manifestation of scattering.

Of course, physically this weak scattering effect is of lesser importance than the possibility of energy exchange between the waves of both families at the singular point.

We will now study the case $D = 2$ (two physical space variables). I do not know whether the elliptic case ('+' in the normal form of the singularity of the light hypersurface) is possible for generic hyperbolic variational systems. We will study the hyperbolic case ('−' in the normal form), which is clearly possible.

After a trivial change of variables, the normal form can now be written as

$$H = p_1 q_1 + p_2^2, \quad dz = \frac{p_1\,dq_1 - q_1\,dp_1 + p_2\,dq_2 - q_2\,dp_2}{2}. \tag{1}$$

The equations of the characteristics of the hypersurface $H = 0$ are

$$\dot{p}_1 = -p_1, \quad \dot{q}_1 = q_1, \quad \dot{p}_2 = 0, \quad \dot{q}_2 = -2p_2, \quad \dot{z} = 0 \tag{2}$$

(where the dot means the derivative with respect to an auxiliary variable t along the characteristics; this variable t does not coincide with the time function $T(p, q, z)$ with respect to which the system is hyperbolic).

A momentary wave front in the physical plane is represented in the contact 5-manifold by an integral curve of the contact structure which belongs to the light hypersurface $H = 0$. The codimension of the incoming separatrix $q_1 = 0$ in the light hypersurface is equal to 1. Hence a generic wave front passes through the singularity at isolated time moments at isolated points.

Assume that the integral curve representing the momentary front transversally intersects the incoming separatrix at the point with coordinates $p_1 = 1$, $p_2 = q_1 = q_2 = z = 0$ (the general case can be reduced to this one by an obvious change of variables). Consider the characteristics of the light hypersurface that issue from the points of our curve. They form a 2-dimensional Legendre submanifold of the light hypersurface (this Legendre surface is formed by the contact elements that are tangent to the big front generated by our momentary front in space–time; the big front is the surface sweept by the momentary fronts in the $(1 + 2)$-dimensional space–time).

Lemma. *The above Legendre surface is reducible to the normal form*

$$p_1 = e^{-t}, \quad p_2 = u, \quad q_1 = -u^2 e^t, \quad q_2 = 2ut, \quad z = -\frac{u^2}{2}, \tag{3}$$

by a contactomorphism preserving the light hypersurface $H = 0$ (u, t are parameters along the surface).

Proof. Consider the intersection of the Legendre surface and the hypersurface $p_1 = 1$. This curve is transversal to the incoming separatrix. In a neighborhood of our point, the local coordinates on the hypersurface (1) are (p_1, p_2, q_2, z). In these coordinates, the equation of the incoming separatrix is $p_2 = 0$. Hence $dp_2 \neq 0$ along the curve. Thus, the restriction of p_2 to the curve can be chosen as a parameter u along this curve:

$$p_1 = 1, \quad p_2 = u, \quad q_1 = -u^2, \quad q_2 = f(u), \quad z = g(u).$$

The symplectomorphism $(p_1, p_2, q_1, q_2) \mapsto (p_1, p_2, q_1, q_2 - f(p_2))$ kills f, preserving H. The corresponding contactomorphism reduces g to the form $-u^2/2$, since along our integral curve of the contact structure, $dz = p_1 \, dq_1/2 = -u \, du$. Thus, our parametrised curve is reduced to the form

$$p_1 = 1, \quad p_2 = u, \quad q_1 = -u^2, \quad q_2 = 0, \quad z = -\frac{u^2}{2}.$$

The characteristics of our hypersurface $H = 0$ that issue from the points of this curve sweep the Legendre surface described in the lemma (by the equations (2) of the characteristics).

Now we will use the normal form (3) to describe the singularity of the big front in space–time, and the perestroika of the momentary fronts. The big front is the projection of the Legendre surface (3) from 5-space to 3-space along the fibers of the

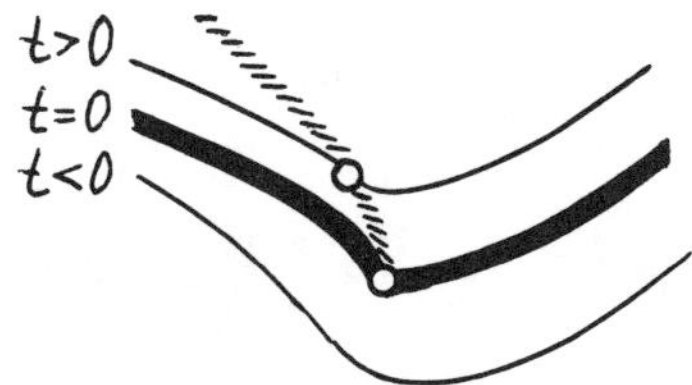

Figure 122: The perestroika of momentary fronts moving in the plane

projective cotangent bundle. In contrast with the case of one space dimension, in the case of two space dimensions there is no visible distinction between generic projections and generic Legendre projections Hence below I will describe the singularities of generic projections of the surface (3) to 3-dimensional space–time.

The final result is the following description of the generic perestroikas of momentary fronts (Fig. 122).

The momentary front, which was a smooth curve before the moment of perestroika, acquires a singular point. At the moment of perestroika the front forms an angle. Both branches of this broken front have a tangent at the singular point (distinct from each other), but the curvatures are infinite. The equation of each branch at the singular point can be written as $y = x/\ln x^2 + o(x/\ln x^2)$.

After the moment of perestroika the front has a continuous tangent at each point. However, it has a singular point, at which the curvature is infinite. The equation of the front in a neighborhood of the singular point can be written as $y = x/\ln x^2 + o(x/\ln x^2)$.

In order to derive these results, we introduce new parameters (A, B) on the surface (3), defining them using the twofold 'covering' of a halfcone by the plane:

$$p_1 = A^2, \quad -q_1 = B^2, \quad p_2 = AB. \tag{4}$$

Then $u = AB$ and $t = -\ln A^2$. Hence the surface (3) is now parametrised by the equations

$$p_1 = A^2, \quad p_2 = AB, \quad q_1 = -B^2, \quad q_2 = -2AB\ln A^2, \quad z = -\frac{A^2 B^2}{2}. \tag{5}$$

The 3 coordinates of the projection of this surface to 3-space are even functions of (A, B). They can be written as the result of substituting the 5 functions (5) in the smooth functions $F_i(p_1, p_2, q_1, q_2, z)$.

We can choose the origin in space–time at the singular point. The functions F_i vanish at the origin.

Consider the Taylor series of F_i at the origin. By substitution of (5), we obtain a quasi-Taylor series, whose quasimonomials are of the form $A^k B^l (\ln A^2)^m$, $m \le k$. In the natural hierarchy of quasimonomials, by their order at the origin, a quasimonomial of this type is neglectable with respect to the quasimonomials with smaller $k + l$, and

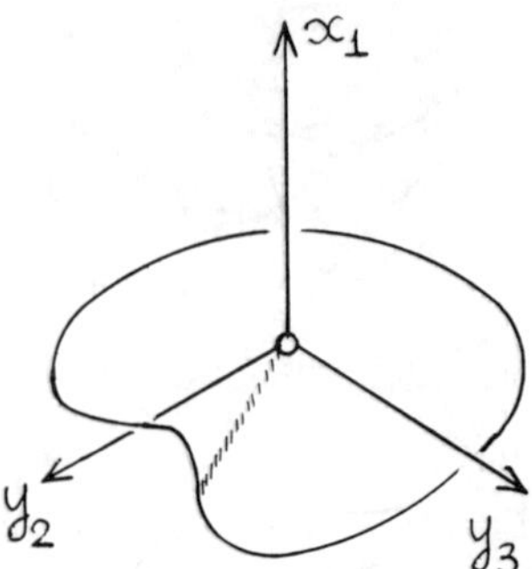

Figure 123: The singular ray on the big front in space–time (elliptic case)

the quasimonomials with larger $k+l$ are neglectable with respect to this monomial. For two quasimonomials with equal $k+l$, the one having largest m is larger (its order of vanishing is smaller).

Taking this into account, we find that the principal parts of the F_i at the origin are proportional to $AB \ln A^2$. The coefficient at this quasimonomial is a vector in our 3-space. We can choose coordinates in 3-space such that the first coordinate axis will contain this vector. This reduces our projection to the form

$$\begin{aligned}
x_1 &= AB \ln A^2 + \cdots, \\
x_2 &= P(A,B) + \cdots, \\
x_3 &= Q(A,B) + \cdots,
\end{aligned} \tag{6}$$

where P and Q are quadratic forms and the dots indicate terms of higher orders.

The quadratic forms P and Q define a quadratic map from the (A,B)-plane to the (x_2, x_3)-plane. A generic quadratic map from one plane to another is reducible to one of the two normal forms

$$y_2 = C^2 - D^2, \quad y_3 = 2CD \qquad \text{(elliptic case)},$$
$$y_2 = C^2, \qquad\quad y_3 = D^2 \qquad \text{(hyperbolic case)} \tag{7}$$

by a linear change of variables

$$(A,B) \mapsto (C,D), \qquad (x_2, x_3) \mapsto (y_2, y_3).$$

Neglecting higher order terms in (6), we obtain in the elliptic case the representation of the projection as the graph of the function $AB \ln A^2$ of the variables y_2, y_3 (Fig. 123). This function is smooth, except on the ray corresponding to the line $A = 0$, on which the tangent plane is vertical and the curvature is (logarithmically) infinite.

In the hyperbolic case the 'function' $AB \ln A^2$ of the variables y_2, y_3 is defined only on the positive quadrant ($y_2 \geq 0$, $y_3 \geq 0$), and has two values for each interior point.

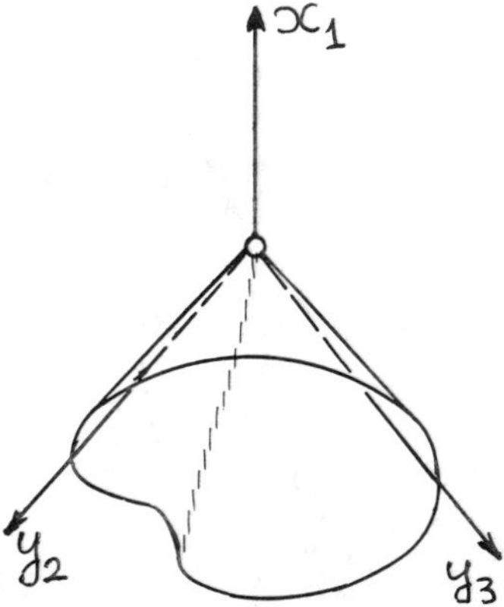

Figure 124: The singular ray on the big front in space–time (hyperbolic case)

Still, the graph has one singular ray (with singularity of the same, logarithmic, type as in the elliptic case, Fig. 124).

Taking into account the neglected terms, the topological picture remains unchanged: the projected surface is still homeomorphic to the plane and still has a singular ray, along which the tangent plane is continuous but the curvature of the transversal section becomes infinite.

In order to find the pattern of the perestroika of the front we consider a generic smooth time function that vanishes at the origin of our 3-space.

The first terms in the quasi-Taylor expansion of this function at the origin are:

$$T = aAB \ln A^2 + bA^2 + cAB + dB^2 + \cdots.$$

The zero level lines on the (A, B)-plane have the asymptotics

$$B = -\frac{b}{a} \frac{A}{\ln A^2} + \cdots,$$

$$A = -\frac{d}{a} \frac{B}{\ln B^2} + \cdots. \tag{8}$$

Hence the level lines of the time function on the (A, B)-plane are similar to those of the function AB, the two separatrices corresponding to the zero level being slightly deformed (Fig. 125).

Remark. The characteristics of the light hypersurface belonging to our Legendre surface are represented on the (A, B)-plane by curves $AB = \text{const}$.

For a system of partial differential equations that is hyperbolic with respect to the given time function T, the characteristics of the light hypersurface are transversal to the isochrones. Hence, for the Legendre surface and time functions describing the propagation of wave fronts for a hyperbolic system of partial differential equations, the curves $T(A, B) = \text{const}$ should not touch the hyperbolas $AB = \text{const}$. This causality condition implies a restriction on the coefficients of the formulas (8). In the general

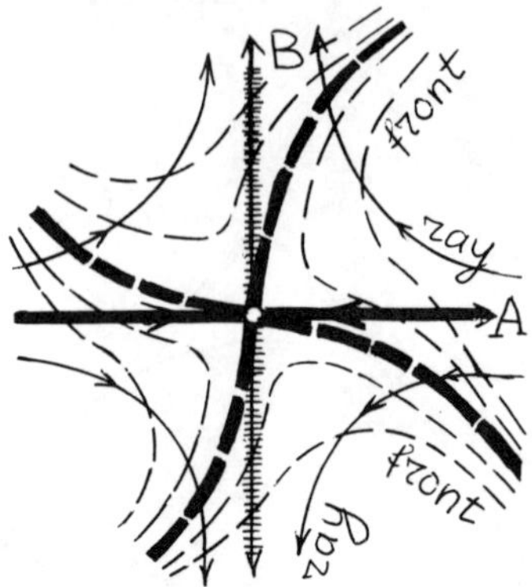

Figure 125: The rays and momentary fronts on the (A, B)-plane

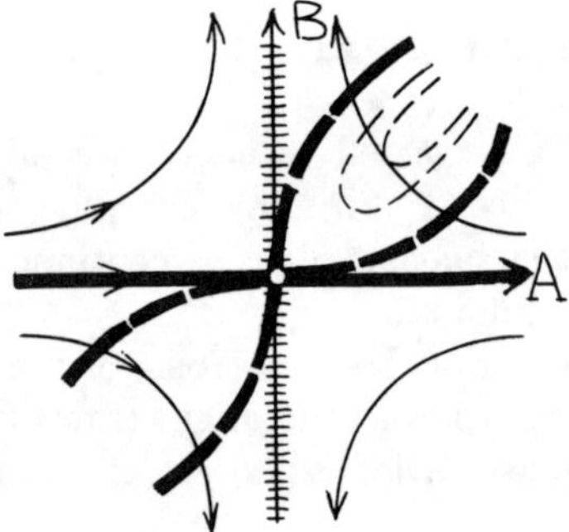

Figure 126: An impossible configuration of rays and fronts for hyperbolic systems

problem of contact geometry, which is not related to the theory of partial differential equations, a pattern as in Fig. 126 is also possible.

Finally, in order to obtain the sweeping of the big front by the momentary fronts, we project the (A, B)-plane, with its characteristics $AB = $ const and isochrones $T(A, B) = $ const, onto the big front, by the projection defined by (6), (7). An easy computation proves

Lemma. *The quadratic maps (7) send central conics (defined by $fC^2 + gCD + hD^2 = $ const) to conics.*

Hence the families of projections of the characteristics from the (A, B)-plane to the (y_2, y_3)-plane are the families of conics shown in Fig. 127.

Remark. The hyperbolic case splits into 2 subcases, since the lines $A = 0$, $B = 0$ may be separated by the lines $C = 0$, $D = 0$ or may be not separated by them on the

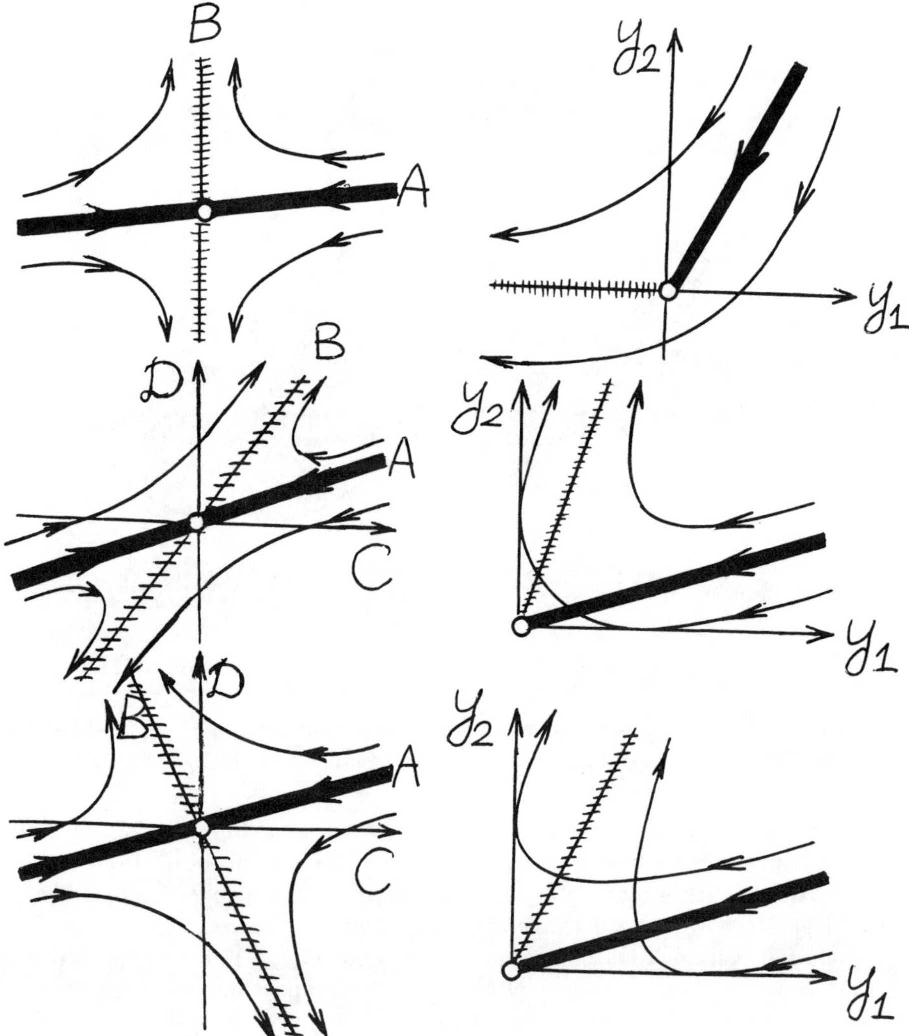

Figure 127: The projection of the characteristics to the (y_2, y_3)-plane

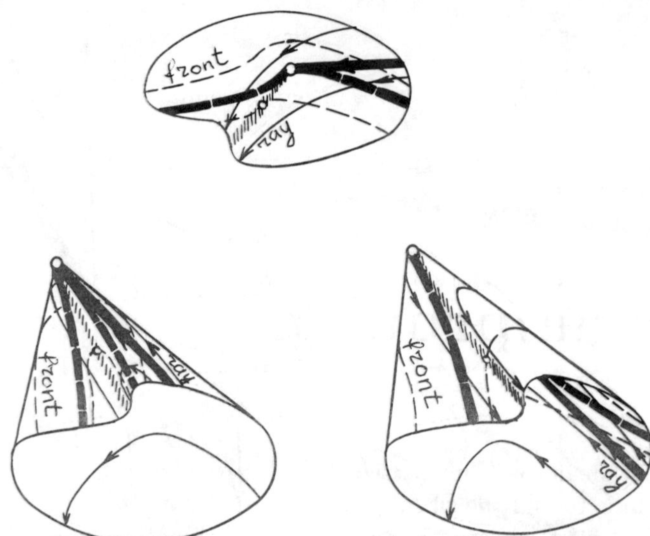

Figure 128: The rays and fronts on the big front

projective line.

Finally we lift the pattern of wave fronts and rays from the (y_1, y_2)-plane to the big front (Fig. 128) and obtain the required perestroika pattern of Fig. 122.

Remark. The typical singularity of the big front in 3-dimensional space–time is probably diffeomorphic to the typical singularity of the momentary front in physical 3-space. Hence such a momentary front should have (at a generic moment of time) singular points, at which the (logarithmically) singular lines end.

An initially generic smooth front moving in a nonhomogeneous 3-space will acquire singularities at a certain moment of time. At the moment following the birth of the singularity, the front must have 2 isolated singular points of nonstrict hyperbolicity, at which the transformation of waves occurs. These points are connected by a singular curve (representing those rays that have earlier experienced the transformation).

Bibliography

[1] V.I. Arnold, Wave front evolution and equivariant Morse Lemma, *Comm. Pure Appl. Math.* **29** no. 6 (1976), 557–582. (Page(s): xi, 32, 57, 58, 64, 128, 171, 173.)

[2] V.I. Arnold, Normal forms for functions near degenerate critical points, the Weyl groups A_k, D_k, E_k and Lagrangian singularities, *Funct. Anal. Appl.* **6** (1972), 254–272. (Page(s): xi, 19, 20, 52, 129.)

[3] V.I. Arnold, Critical points of functions on a manifold with boundary, the simple Lie groups B_k, C_k, F_4 and singularities of evolutes, *Russian Math. Surveys* **33** no. 5 (1978), 99–116. (Page(s): xii, 69, 102, 129, 134, 136, 138.)

[4] O.V. Ljashko, Classification of critical points of functions on a manifold with singular boundary, *Funct. Anal. Appl.* **17** (1983), 187–193. (Page(s): xii, 157, 171, 207.)

[5] O.P. Shcherbak, Singularities of families of evolvents in the neighbourhood of an inflection point of the curve, and the group H_3, generated by reflections, *Funct. Anal. Appl.* **17** (1983), 301–303. (Page(s): xii, 157, 171, 207.)

[6] O.P. Shcherbak, H_4 in the obstacle bypassing problem, *Uspekhi Mat. Nauk* **39** no. 4 (1984), 114. (In Russian.) (Page(s): xii, 199.)

[7] V.I. Arnold, Singularities in the calculus of variations, *Uspekhi Mat. Nauk* **39** no. 5 (1984), 256. (In Russian.) (Page(s): xii.)

[8] A.B. Givental, Singular Lagrange varieties and their Lagrange mappings, *Itogi Nauk i Tekhn. VINITI Sovr. Probl. Mat. Nov. Dost.* **33** (1988), 55–112. (In Russian.) (English translation to appear in: J. Soviet Math.) (Page(s): xii, 7, 81, 120, 157, 187, 189, 192, 201, 203, 204, 209, 212, 229.)

[9] O.A. Platonova, Projections of smooth surfaces, *Tr. Sem. I.G. Petrovskii* **10** (1984), 135–149. (In Russian.) (English translation in: *J. Soviet Math.* **35** no. 6 (1986), 2796–2808.) (Page(s): xii, 124–126.)

[10] D. Hilbert, Ueber die vollen Invariantensysteme, *Math. Ann.* **42** (1893), 313–373. (Page(s): 6, 179.)

[11] A.B. Givental, Manifolds of polynomials, having a root of fixed comultiplicity, and the generalized Newton equation, *Funct. Anal. Appl.* **16** (1982), 10–14. (Page(s): 6, 157, 178, 180, 182.)

[12] A.B. Givental, Lagrangian varieties with singularities and irreducible sl_2-modules, *Russian Math. Surveys* **38** no. 6 (1983), 121–122. (Page(s): 6, 157.)

[13] V.I. Arnold, Catastrophe theory, Znanie, Moscow, 1981. (In Russian.) (English translation: Springer, 1986.) (Page(s): 7, 45, 124, 175.)

[14] V.I. Arnold, A.B. Givental, Symplectic geometry, *Itogi Nauk i Tekhn. VINITI* **4** (1985), 5–139. (In Russian.) (English translation in: V.I. Arnold and P.S. Novikov (eds.), Dynamical Systems, Encycl. of Math. Sci. vol IV, Springer, 1980, pp. 4–136.) (Page(s): 7, 44–46, 95.)

[15] A. Weinstein, *Lectures on symplectic manifolds*, Reg. Conf. Ser. Math. vol. 29, Aner. Math. Soc., 1977. (Page(s): 7.)

[16] J. Martinet, Sur les singularités des formes différentielles, *Ann. Inst. Fourier (Grenoble)* **20** no. 1 (1970), 95–178. (Page(s): 11.)

[17] V.I. Arnold, *Supplementary chapters of the theory of ordinary differential equations*, Nauka, Moscow, 1978. (In Russian.) (English translation: *Geometrical methods in the theory of ordinary differential equations*, Grundlehren vol. 250, Springer, 1988.) (Page(s): 11, 30, 229, 230.)

[18] R. Melrose, Equivalence of glancing hypersurfaces, *Invent. Math.* **37** (1976), 165–191. (Page(s): 13, 157, 161.)

[19] M. Kashiwara, T. Kawai, Hidden boundary conditions, *Sûrikaiseki-kenkyûsho Kôkyûroku* **227** (1975), 39–41. (In Japanese.) (Page(s): 13.)

[20] T. Oshima, On analytic equivalence of glancing hypersurfaces, *Papers of the Col. of Gen. Education, Univ. Tokyo* **28** (1978), 51–57. (In Japanese.) Local equivalence of differential forms and their deformations, *Sûrikaiseki-kenkyûsho Kôkyûroku* **266** (1976), 108–129. (In Japanese.) (Page(s): 13.)

[21] A.A. Kirillov, Geometrical quantization, *Itogi Nauk i Tekhn. VINITI* **4** (1985), 162. (In Russian.) (English translation in: V.I. Arnold and P.S. Novikov (eds.), Dynamical Systems, Encycl. of Math. Sci. vol IV, Springer, 1980, pp. 137–172.) (Page(s): 13.)

[22] R. Thom, Topological models in biology, *Topology* **8** no. 3 (1969), 313–335. (Page(s): 18.)

[23] V.I. Arnold, Lagrangian manifold singularities, asymptotic rays and the open swallowtail, *Funct. Anal. Appl.* **15** (1981), 235–246. (Page(s): 19, 156, 157, 162–164, 185, 214.)

[24] O.A. Platonova, Singularities in the problem of the quickiest bypassing of an obstacle, *Funct. Anal. Appl.* **15** (1981), 147–148. (Page(s): 19, 157.)

[25] E.E. Landis, Tangential singularities, *Funct. Anal. Appl.* **15** (1981), 103–114. (Page(s): 19, 125, 157.)

[26] C. McCrory, T. Shifrin, Cusps of the projective Gauss map, *J. Diff. Geom.* **19** (1984), 257–276. (Page(s): 19, 127.)

[27] V.M. Zakalyukin, On Lagrange and Legendre singularities, *Funct. Anal. Appl.* **10** (1976), 23–31. (Page(s): 19.)

[28] V.I. Arnold, S.M. Gusein-Zade, A.N. Varchenko, *Singularities of diferentiable maps*, vol. 1, Nauka, Moscow, (1982), (In Russian.) (English translation: Birkhäuser, 1985.) (Page(s): 19, 20, 34, 55, 60, 65, 66, 70, 77, 112, 132, 147.)

[29] V.M. Zakalyukin, Perestroika of fronts and caustics, depending on parameters, and versality of mappings, *Itogi Nauk i Tekhn. VINITI Sovr. Probl. Mat.* **22** (1983), 56–93. (In Russian.) (English translation: *J. Soviet Math.* **27** (1984), 2713–2735.) (Page(s): 19, 55, 141.)

[30] A.N. Varchenko, Theorem on the topological equisingularity of families of algebraic varieties and polynomial mappings, *Math. USSR Izv.* **36** (1972), 957–1019. (Page(s): 19.)

[31] C.G. Gibson, K. Wirtmüller, A. du Plessis, E.J.N. Looijenga, *Topological stability of smooth mappings*, Lecture Notes in Math. vol. 552, Springer, 1976. (Page(s): 19.)

[32] Y. Colin de Verdière, Nombre des points entiers dans une famille homothetique des domaines de $\mathbf{R}^n$, *Ann. Sci. Ecole Norm. Super. Sér. 4* **10** no. 4 (1977), 554–575. (Page(s): 22, 28, 29.)

[33] A.N. Varchenko, Newton polyhedra and estimates of oscillating integrals, *Funct. Anal. Appl.* **10** (1976), 75–196. (Page(s): 23.)

[34] V.I. Arnold, Index of a singular point of a vector field, Petrovsi—Oleinik inequalities and mixed Hodge structures, *Funct. Anal. Appl.* **12** (1978), 1–11. (Page(s): 23, 25.)

[35] A.G. Hovansky, Geometry of formulas, *Math. Phys. Review* **4** (1984), 1-92. (In: V.I. Arnold, A.N. Varchenko, A.B. Givental, A.G. Hovansky, Singularities of functions, wave fronts, caustics and multidimensional integrals, Harwood, 1984.) (Page(s): 23.)

[36] A.G. Hovanski, Newton polyhedra (resolution of singularities), *Itogi Nauk i Tekhn. VINITI Sovr. Probl. Mat.* **22** (1983), 207–239. (In Russian.) (English translation: *J. Soviet Math.* **27** (1984), 2812–2833.) (Page(s): 23.)

[37] I.M. Vinogradov, *The method of trigonometric sums in the theory of numbers*, Interscience, 1954. (Translated from the Russian.) (Page(s): 24.)

[38] V.N. Karpushkin, Uniform estimation of oscillating integrals in $\mathbf{R}^2$, *Sov. Math. Dokl.* **254** no. 1 (1980), 28–31. (Page(s): 24.)

[39] V.N. Karpushkin, Theorems on uniform estimation of oscillatory integrals with phase depending on two variables, *Tr. Sem. I.G. Petrovskii* **10** (1984), 150-169. (In Russian.) (English translation:*J. Soviet Math.* **35** no. 6 (1986), 2809–2826.) (Page(s): 24.)

[40] F. Pham, Caustics and microfunctions, *Preprint Dept. des Math. Univ. Nice* (1976). (Page(s): 25.)

[41] V.A. Vasilev, The asymptotics of exponential integrals in the complex domain, *Funct. Anal. Appl.* **13** (1979), 239–247. (Page(s): 25.)

[42] A.N. Varchenko, Gauss–Manin connection of isolated singular points and Bernstein polynomial, *Bull. Sci. Math. Sér. 2* **104** (1980), 205–223. (Page(s): 25, 73.)

[43] A.N. Varchenko, Asymptotic mixed Hodge structure in vanishing cohomology, *Math. USSR Izv.* **18** no. 3 (1982), 469–512. (Page(s): 25, 73.)

[44] V.I. Arnold, S.M. Gusein-Zade, A.N. Varchenko, *Singularities of diferentiable maps*, vol. 2, Nauka, Moscow, 1984. (In Russian.) (English translation: Birkhäuser, 1988.) (Page(s): 25, 27, 73, 138.)

[45] J.H.M. Steenbrink, *Mixed Hodge structure on the vanishing cohomology*, In: Real and Complex Singularities, Nordic Summer School Oslo, 1976, Sijthoff & Noordhoff, 1977, pp. 523–563. (Page(s): 25.)

[46] A.N. Varchenko, On the semicontinuity of spectra and estimates from above of the number of singular points of a projective hypersurface, *Sov. Math. Dokl.* **270** no. 6 (1983), 1294–1297. (Page(s): 25.)

[47] J.H.M. Steenbrink, Semicontinuity of the singular spectrum, *Invent. Math.* **79** no. 3 (1985), 557-566. (Page(s): 25.)

[48] A.N. Varchenko, The complex singular index does not change along the stratum $\mu = \mathrm{const}$, *Funct. Anal. Appl.* **16** (1982), 1–9. (Page(s): 25.)

[49] A.N. Varchenko, An estimate from below of the codimension of the stratum $\mu = \mathrm{const}$ by mixed Hodge structures, *Vestnik Moskov. Univ. Ser. Mat. Mekh.* no. 6 (1982), 28–31. (In Russian.) (English translation: Moscow Univ. Math. Bull.) (Page(s): 25.)

[50] V.I. Arnold, On some problems in singularity theory, *Tr. Sem. S.L. Sobolev po kubaturn. formulam, Novosibirsk* no. 1 (1976), 5–15. (In Russian.) (English translation in: Singularities, Proc. Symp. Pure Math. vol. 40^1, Amer. Math. Soc., 1983, pp. 57–69.) (Page(s): 26, 106.)

[51] D.A. Gudkov, Topology of real projective algebraic varieties, *Russian Math. Surveys* **29** no. 3 (1974), 3–79. (Page(s): 26.)

[52] V.M. Harlamov, *Topology of real algebraic varieties*, In: I.G. Petrovsky, Selected Work, Systems of Partial Differential Equations, Algebraic Geometry, Nauka, Moscow, 1986, pp. 465–493. (In Russian.) (Page(s): 26.)

[53] O.Ja. Viro, Progress in the topology of real algebraic varieties in the last six years, *Moscow Univ. Math. Bull.* **41** no. 3 (1986), (Page(s): 26.)

[54] Ju.V. Chekanov, Asymptotics of the number of maxima for a product of linear functions of two variables, *Moscow Math. Bull.* **41** (1986), 85–87. (Page(s): 26.)

[55] Z. Füredi, I. Patasti, Arrangements of lines with a large number of triangles, *Proc. Amer. Math. Soc.* **92** (1984), 561–566. (Page(s): 26.)

[56] B. Grünbaum, *Arrangements and spreads*, Amer. Math. Soc, 1965. (Page(s): 26.)

[57] A. Durfee, N. Kronefeld, H. Munson, J. Roy, I. Westby, Critical points of real polynomials in two variables, *Preprint Mount Hlyoke Coll.* (1989), (Page(s): 26.)

[58] Y. Miyaoka, The maximal numbers of quotient singularities on surfaces with given numerical invariants, *Math. Ann.* **268** (1984), 159–171. (Page(s): 27.)

[59] A.N. Varchenko, Asymptotics of integrals and Hodge structures, *Itogi Nauk i Tekhn. VINITI Sovr. Probl. Mat.* **22** (1983), 130–166. (In Russian.) (English translation: *J. Soviet Math.* **27** (1984), 2760-2784.) (Page(s): 27.)

[60] A.N. Varchenko, Number of lattice points in families of homothetical domains in $\mathbf{R}^n$, *Funct. Anal. Appl.* **17** (1983), 79–83. (Page(s): 28, 29.)

[61] B. Randol, On the asymptotic behaviour of the Fourier transform of the indicator function of a convex set, *Trans. Amer. Math. Soc.* **139** (1969), 279–285. (Page(s): 28, 29.)

[62] G.E. Andrews, A lower bound for the volumes of strictly convex bodies with many boundary points, *Trans. Amer. Math. Soc.* **106** (1965), 270-279. (Page(s): 30.)

[63] S.M. Konjagin, S. Sevastjanov, Estimation of the number of vertices of a convex integer polyhedron by its volume, *Funct. Anal. Appl.* **18** (1984), (Page(s): 30.)

[64] V.I. Arnold, Statistics of convex polygons with integer vertices, *Funct. Anal. Appl.* **14** no. 2 (1980), 1–3. (Page(s): 30.)

[65] A.S. Pyartli, Diophantine approximation on submanifolds of euclidean space, *Funct. Anal. Appl.* **3** no. 4 (1969), 59–62. (Page(s): 30.)

[66] V.G. Sprindzuk, *Mahler's problem in metric number theory*, Amer. Math. Soc., 1969. (Translated from the Russian.) (Page(s): 30.)

[67] Ju.S. Il'jashenko, A condition for the steepness of analytic functions, *Russian Math. Surveys* **41** no. 1 (1986), 193–194. (Page(s): 30.)

[68] E.E. Landis, Uniform steepness exponents, *Uspekhi Mat. Nauk* **41** no. 4 (1986), 179. (In Russian.) (Page(s): 30.)

[69] Ja.B. Zeldovich, Gravitational instability: an approximate theory for large density perturbations, *Astron. Astrophysics* **5** (1970), 84–89. (Page(s): 31, 34.)

[70] V.I. Arnold, On the Newtonian attraction of the clusters of dust-like matter, *Uspekhi Mat. Nauk* **37** no. 4 (1982), 125. (In Russian.) (Page(s): 31.)

[71] A.A. Roitvarf, On a two-valued velocity field with a root singularity, *Moscow Univ. Math. Bull.* no. 3 (1987), 40-43. (Page(s): 31.)

[72] V.I. Arnold, Perestroikas of singularities of potential flows in collisionless media and metamorphoses of caustics in 3-space, *Tr. Sem. I.G. Petrovskii* **8** (1982), 21–57. (In Russian.) (English translation: *J. Soviet Math.* **32** no. 3 (1986), 229–257.) (Page(s): 32, 119, 157, 173–175, 178.)

[73] V.I. Arnold, Catastrophe theory, *Priroda* **10** (1979), 63–84. (In Russian.) (Page(s): 32.)

[74] V.I. Arnold, S.F. Shandarin, Ja.B. Zeldovich, The large-scale structure of the universe I, *Geophys. Astrophys. Fluid Dynam.* **20** (1982), 111–130. (Page(s): 32, 34.)

[75] S.F. Shandarin, Percolation theory and the celular-netlike structure of the Universe, *Preprint Keldysh Inst. Appl. Math.* **137** (1982), (In Russian.) (Page(s): 34.)

[76] S.F. Shandarin, A. Dorochkevich, Ja.B. Zeldovich, Large scale structure of the Universe, *Uspekhi Fiz. Nauk* **139** (1983), 83–134. (In Russian.) (English translation: *Soviet Phys. Uspekhi* **26** no. 1 (1983), 46–83.) (Page(s): 34.)

[77] J.F. Nye, J.H. Hannay, The orientation and distortions of caustics in geometrical optics, *Optica Acta* **31** no. 1 (1984), 115–130. (Page(s): 35.)

[78] Ju.V. Chekanov, Caustics in geometrical optics, *Funct. Anal. Appl.* **20** (1986), 223–226. (Page(s): 35.)

[79] J. Guckenheimer, Caustics and nondegenerate hamiltonians, *Topology* **13** no. 2 (1974), 127–133. (Page(s): 35.)

[80] V.I. Arnold, V.A. Vasil'ev, V.V. Gorjunov, O.V. Ljashko, Singularities 2, *Itogi Nauk i Tekhn. VINITI Sovr. Probl. Mat. Fund. Napravl.* **39** (1989), (In Russian.) (English translation: Encycl. of Mat Sci. vol. 39, Springer.) (Page(s): 40, 102, 105, 147.)

[81] V.I. Arnold, Catastrophe theory, *Itogi Nauk i Tekhn. VINITI Sovr. Probl. Mat. Fund. Napravl.* **5** (1986), 219–277. (English translation: Dynamical Systems 5, Encycl. of Mat Sci. vol. V, Springer, 1990.) (Page(s): 40.)

[82] I.A. Bogaevsky, Perestroikas of singularities of minima functions and bifurcations of shock waves for the Burgers equation with vanishing viscosity, *Alegbra and Analysis* **1** no. 4 (1989), 1–16. (Page(s): 40.)

[83] A.R. Forsyth, *Theory of differential equations*, vol. 6, Cambridge Univ. Press, 1906. pp. 101; Chapt. XIII, no. 207. (Page(s): 41.)

[84] V.A. Florin, Some simplest nonlinear problems of consolidation of watersaturated soils, *Izv. Akad. Nauk SSSR Otd. Tekhn. Nauk* **9** (1948), 1389–1397. (In Russian.) (Page(s): 41.)

[85] S.N. Gurbatov, A.I. Saichev, I.G. Jakushkin, Nonlinear waves and one-dimensional turbulence in nondispersive media, *Soviet Phys. Uspekhi* **26** no. 10 (1983), 857–876. (Page(s): 42.)

[86] V.I. Arnold, First steps of symplectic topology, *Russian Math. Surveys* **41** no. 6 (1986), 1–21. (Page(s): 42.)

[87] M.L. Bialyi, L.V. Polterovich, Geodesical flows on a two-dimensional tore and phase transitions commensurability-uncommensurability, *Funct. Anal. Appl.* **20** (1986), 260–266. (Page(s): 42.)

[88] M.L. Bialyi, L.V. Polterovich, Lagrangian singularities of invariant tori of hamiltonian systems with two degrees of freedom, *Invent. Math.* **97** no. 2 (1989), 291–303. (Page(s): 42.)

[89] L.V. Polterovich, Strongly optical Lagrange manifolds, *Math. Notes* **45** no. 2 (1989), 95–104. (Page(s): 42.)

[90] C.G.J. Jacobi, *Vorlesungen über Dynamik*, G. Reimer, 1884. (Reprint: Chelsea, 1969.) (Page(s): 42, 82.)

[91] V.I. Arnold, Ramified covering $CP^2 \to S^4$, hyperbolicity and projective topology, *Sib. Math. J.* **29** no. 5 (1988), 36–47. (Page(s): 42.)

[92] V.I. Arnold, On functions with mild singularities, *Funct. Anal. Appl.* **23** no. 3 (1989), 1–10. (Page(s): 42, 80, 109, 111.)

[93] V.I. Arnold, *Mathematical methods of classical mechanics*, Nauka, Moscow, 1974. (In Russian.) (English translation: Graduate Texts in Math. vol. 60, Springer, 1989.) (Page(s): 44, 161.)

[94] V.I. Arnold, Contact manifolds, Legendre mappings and singularities of wave fronts, *Uspekhi Mat. Nauk* **29** no. 4 (1974), 153–154. (In Russian.) (Page(s): 55.)

[95] V.M. Zakaljukin, Bifurcations of wave fronts, depending on one parameter, *Funct. Anal. Appl.* **10** (1976), 139–140. (Page(s): 60, 65, 86.)

[96] V.I. Bakhtin, Topologically normal forms of the perestroikas of the series D, *Vestnik Moskov. Mat. Bull.* **1** no. 4 (1987), 58–61. (In Russian.) (English translation: Moscow Univ. Math. Bull. (1987).) (Page(s): 60.)

[97] V.I. Arnold, *Critical points of smooth functions*, In: Proc. Internat. Congress Math. Vancouver, 1974, vol. 1, Canad. Math. Congress, 1975, pp. 19–39. (Page(s): 60.)

[98] V.I. Arnold, Indices of singular points of 1-forms on manifolds with boundary, convolution of invariants of groups, generated by reflections, and singular projections of smooth hypersurfaces, *Russian Math. Surveys* **34** no. 2 (1979), 1–42. (Page(s): 60, 63, 64, 69, 71, 72, 128, 129, 141, 143, 208.)

[99] L. Solomon, Invariants of finte reflction groups, *Nagoya Math. J.* **22** (1963), 57–64. (Page(s): 63.)

[100] A.B. Givental, Convolution of invariants of groups, generated by reflections, and related to simple singularities of functions, *Funct. Anal. Appl.* **14** (1980), 81–89. (Page(s): 64, 69.)

[101] T. Yano, Flat coordinate system for the deformation of type E_6, *Proc. Japan Acad.* **57A** (1981), 413–414. (Page(s): 65.)

[102] K. Saito, On the periods of primitive integrals I, *RIMS* (1982), 1–235. (Page(s): 65.)

[103] D. Eisenbud, H. Levin, An algebraic formula for the degree of C^∞ map germ, *Ann. Math.* **106** no. 1 (1977), 19–38. (Page(s): 70.)

[104] A.B. Givental, A.N. Varchenko, Period mapping and intersection form, *Funct. Anal. Appl.* **16** (1982), 83–93. (Page(s): 73, 86, 192.)

[105] E. Looijenga, Homogeneous spaces associated to certain semiuniversal deformations, In: Proc. Intern. Congress Math. Helsinki, 1978, vol. 2, 1980, pp. 529--536. (Page(s): 85.)

[106] V.I. Arnold, Lagrange and Legendre cobordisms I, *Funct. Anal. Appl.* **14** (1980), 167–177. (Page(s): 90 93, 94.)

[107] V.I. Arnold, Lagrange and Legendre cobordisms II, *Funct. Anal. Appl.* **14** (1980), 252–260. (Page(s): 90, 94–96.)

[108] R.O. Wells, Compact real submanifolds of a complex manifold with nondegenerate tangent bundles, *Math. Ann.* **179** (1969), 123–129. (Page(s): 93.)

[109] M. Gromov, Pseudo-holomorphic curves in symplectic manifolds, *Invent. Math.* **82** (1985), 307–347. (Page(s): 93.)

[110] V.B. Tvorogov, Paired integral Fourier operators in the problem of propagation of discontinuities, *VINITI, Moscow* **2872-79** (1979), (In Russian.) (Page(s): 95.)

[111] V.B. Tvorogov, Sharp front and singularities of a class of nonhyperbolic equations, *Dokl. Akad. Nauk SSSR* **244** no. 6 (1979), 1327–1331. (In Russian.) (Page(s): 95.)

[112] J. Ellashberg, Cobordisme des solutions de relations différentielles, In: P. Darzord, N. Desolneux-Moulis (eds.), Sem. Sud-Rhodunien de Geom. Tome 1, Hermann, 1984, pp. 17–32. (Page(s): 95.)

[113] M. Audin, Quelque calculs en cobordisme lagrangien, *Ann. Inst. Fourier (Grenoble)* **35** no. 3 (1985), 159–194. (Page(s): 95.)

[114] M. Audin, *Cobordisme d'immersions Lagrangienne set legendrienne*, Traveaux en Cours 20, Hermann, 1987. (Page(s): 95.)

[115] V.A. Vassilyev [Vasil'ev], *Lagrange and Legendre characteristic classes*, Gordon & Breach, 1988. (Page(s): 95, 98, 100, 102.)

[116] V.A. Vasil'ev, Characteristic classes of Lagrangian and Legendre manifolds dual to singularities of caustics and wave fronts, *Funct. Anal. Appl.* **15** (1981), 164–173. (Page(s): 96, 98, 102.)

[117] V.P. Maslov, *Théorie des perturbations et méthodes asymptotiques*, Dunod, 1972. (Translated from the Russian.) (Page(s): 97.)

[118] V.A. Vasil'ev, Self-intersections of wave fronts and Legendre (Lagrangian) characteristic numbers, *Funct. Anal. Appl.* **16** (1982), 131–133. (Page(s): 100.)

[119] D.B. Fuks, Maslov—Arnold characteristic classes, *Soviet Math. Dokl.* **9** (1968), 96–99. (Page(s): 101.)

[120] I.G. Petrovsky, On the diffusion of waves anf the lacunas for hyperbolic equations, *Mat. Sb.* **17** no. 3 (1945), 289–370. (Page(s): 102.)

[121] V.A. Vasil'ev, Sharpness and the local Petrovskii condition for strictly hyperbolic operators with constant coefficients, *Math USSR Izv.* **28** (1987), 233–273. (Page(s): 102.)

[122] A.N. Varchenko, On normal forms of smoothness of solutions of hyperbolic equations, *Izv. Akad. Nauk. SSSR* **51** no. 3 (1987), 652–665. (In Russian.) (Page(s): 102.)

[123] A.B. Givental, Twisted Picard—Lefshetz formulas, *Funct. Anal. Appl.* **22** (1988), 10–18. (Page(s): 103.)

[124] A.B. Givental, V.V. Shekhtman, Monodromy groups and Hecke algebras, *Uspekhi Mat. Nauk* **12** no. 4 (1987), 138–139. (In Russian.) (Page(s): 103.)

[125] P. Cartier, Dévelopments récents sur les groupes de tresses, applications à la topologie et à l'algèbre, *Sém. Bourbaki* **716** (1989), 1–42. (Page(s): 103.)

[126] E. Brieskorn, Sur les groupes de tresses (d'après V.I. Arnold), *Sèm. Bourbaki* **401** (1971), 1–28. (Page(s): 103.)

[127] V.I. Arnold, Critical points of funcions and the classification of caustics, *Uspekhi Mat. Nauk* **29** no. 3 (1974), 243-244. (In Russian.) (Page(s): 104.)

[128] E. Looijenga, The complement of the bifurcation variety of a simple singularity, *Invent. Math.* **23** no. 2 (1974), 105–116. (Page(s): 104, 105.)

[129] H. Knörrer, Zum $K(\pi, 1)$-Problem für isolierte Singularitäten von vollstandiger Durchschnitten, *Comp. Math.* **45** no. 3 (1982), 333–340. (Page(s): 105, 147.)

[130] V.V. Gorjunov, Projections of the 0-dimensional complete intersections onto the line and the $K(\pi, 1)$ conjecture, *Russian Math. Surveys* **37** no. 3 (1982), 206–208. (Page(s): 105, 133, 147.)

[131] V.V. Gorjunov, Geometry of bifurcations of simple projections onto a line, *Funct. Anal. Appl.* **15** (1981), 77–82. (Page(s): 105, 130, 131, 133, 136, 144, 146.)

[132] V.V. Gorjunov, Bifurcation diagrams of some simple and quasihomogeneous singularities, *Funct. Anal. Appl.* **17** (1983), 97-108. (Page(s): 105.)

[133] V.V. Gorjunov, Singularities of projections of full intersections, *Itogi Nauk i Tekhn. VINITI Sovr. Probl. Mat.* **22** (1983), 167-206. (In Russian.) (English translation: *J. Soviet Math.* **27** (1984), 2785–2811.) (Page(s): 105, 129, 134, 146.)

[134] D.B. Fuks, Cohomology of the braid group mod 2, *Funct. Anal. Appl.* **4** (1970), 143–151. (Page(s): 106.)

[135] G. Segal, Configuration-spaces and iterated loop-spaces, *Invent. Math.* **21** no. 3 (1973), 213–221. (Page(s): 106, 108.)

[136] F.R. Cohen, T.J. Lada, J.P. May, *The homology of iterated loop spaces*, Lecture Notes in Math. vol 533, Springer, 1976. (Page(s): 106.)

[137] V.A. Vasil'ev, Stable cohomology of the complements of the discriminants of the deformations of singularities of smooth functions, *Itogi Nauk i Tekhn. VINITI Sovr. Probl. Mat. Nov. Dost.* **33** (1988), 3–29. (In Russian.) (English translation to appear in: J. Soviet Math.) (Page(s): 106, 107.)

[138] V.A. Vasil'ev, Topology of complements to discriminants and loop spaces, In: Theory of singularities and some applications, Amer. Math. Soc., 1990. (Page(s): 106, 108.)

[139] M.L. Gromov, Ja.M. Eliashberg, Construction of a smooth mapping with a given Jacobian, *Funct. Anal. Appl.* **7** (1973), (Page(s): 114.)

[140] Ja.M. Eliashberg, On the singularities of fold type, *Math. USSR Izv.* (1971), (Page(s): 115.)

[141] A.B. Givental, Lagrange embeddings of surfaces and the open Whitney umbrella, *Funct. Anal. Appl.* **20** (1986), (Page(s): 116, 121.)

[142] V.I. Arnold, Sweeping of caustics by the cuspidal edge of a moving front, *Uspekhi Mat. Nauk* **36** no. 4 (1981), 233. (In Russian.) (Page(s): 119.)

[143] O.P. Shcherbak, Projectively dual curves and Legendre singularities, *Tr. Tbilissk. Univ.* **232–233** no. 13–14 (1982), 280–336. (In Russian.) (English translation: *Selecta Math. Sovietica* **5** (1986), 391–421.) (Page(s): 120, 157, 184, 185.)

[144] O.P. Shcherbak, Wave fronts and reflection groups, *Russian Math. Surveys* **43** no. 3 (1988), 149–194. (Page(s): 120, 157, 174, 184, 199, 200, 201, 209, 214, 217.)

[145] A.A. Davydov, Normal forms of diferential equations, unresolved in derivative, *Funct. Anal. Appl.* **19** no. 2 (1985), 1–10. (Page(s): 120.)

[146] A.A. Davydov, The normal form of slow motions of an equation of relaxation type and fibrations of binomial surfaces, *Math. USSR Sb.* **60** no. 1 (1988), 133–141. (Page(s): 120.)

[147] V.I. Arnold, Implicit differential equations, contact structures and relaxation oscillations, *Uspekhi Mat. Nauk* **40** no. 5 (1985), 188. (See also: Lecture Notes in Math. vol. 1334, Springer, 1988, pp. 173– 179.) (Page(s): 120.)

[148] V.I. Arnold, Singularities of systems of rays, *Russian Math. Surveys* **38** no. 2 (1983), 87–176. (Page(s): 124, 184.)

[149] V.S. Kulikov, The calculus of singularities of incluson of a general algebraic surface in the projective space P^3, *Funct. Anal. Appl.* **17** (1983), 176–186. (Page(s): 127.)

[150] G. Salmon, *A treatise on the analytic geometry of three dimensions*, vol. 2, Chelsea, reprint, 1965. (Page(s): 127.)

[151] C. McCrory, T. Shifrin, R. Valery, *The Gauss map of a generic hypersurface in* $\mathbf{R}^4$, Preprint Athenes, Georgia, USA, 1987. (Page(s): 127.)

[152] A.D.R. Choudary, A. Dimca, On the dual and Hessian mapping of projective hypersurfaces, *Math. Proc. Cambridge Philos. Soc.* **101** (1987), 461–468. (Page(s): 127.)

[153] T.F. Banchoff, T. Gafney, C. McCrory, *Cusps of Gauss mapping*, Research Notes in Math. vol. 55, Pitman, 1982. (Page(s): 129.)

[154] M. Giusti, *Classification des singularits isolées simples d'intersections completes*, In: Singularities, Proc. Symp. Pure Math. vol. 40^1, Amer. Math. Soc. 1983, pp.457–494. (Page(s): 131, 147.)

[155] V.I. Matov, Singularity of the maximum function on a manifold with boundary, *Tr. Sem. I.G. Petrovskogo* **6** (1981), 195–222. (In Russian.) (English translation in: *J. Soviet Math.* **33** no. 4 (1986), 1103–1127.) (Page(s): 134.)

[156] V.I. Matov, Unimodular and bimodular germs of functions on manifolds with boundary, *Tr. Sem. I.G. Petrovskogo* **7** (1981), 174–189. (In Russian.) (English translation in: J. Soviet Math. (1986)) (Page(s): 134.)

[157] I.G. Shcherbak, Duality of boundary singularities, *Russian Math. Surveys* **39** no. 2 (1984), 195–196. (Page(s): 134, 135.)

[158] V.A. Vasil'ev, Topology of spaces of functions, having no complicated singularities, *Funct. Anal. Appl.* **23** no. 4 (1989), 24–36. (Page(s): 112.)

[159] O.V. Ljashko, Geometry of bifurcation diagrams, *Itogi Nauk i Tekhn. VINITI Sovr. Probl. Mat.* **22** (1983), 94–129. (In Russian.) (English translation: *J. Soviet Math.* **27** (1984), 2735–2759.) (Page(s): 141–143, 172, 203.)

[160] V.I. Arnold, V.A. Vasil'ev, V.V. Gorjunov, O.V. Ljashko, Singularities 1. Dynamical Systems, *Itogi Nauk i Tekhn. VINITI Sovr. Probl. Mat. Fund. Napravl.* **6** (1988), (English translation to appear as: Encycl. of Math. Sci. vol. 6, Springer.) (Page(s): 141, 142, 144.)

[161] V.V. Gorjunov, Vector fields and functions on the discriminants of complete intersections and on bifurcation diagrams of projections, *Itogi Nauk i Tekhn. VINITI Sovr. Probl. Mat. Nov. Dost.* **33** (1988), 31–54. (English translation to appear in: J. Soviet Math.) (Page(s): 144.)

[162] D. Siersma, *Isolated line singularities*, In: Singularities, Proc. Symp. Pure Math. vol. 40^2, Amer. Math. Soc., 1983, pp. 485-496. (Page(s): 146.)

[163] R. Melrose, Equivalence of glancing hypersurfaces II, *Math. Ann.* **255** no. 2 (1981), 159–198. (Page(s): 161, 162, 166.)

[164] E.E. Landis, Tangential singularities in contact geometry, *Uspekhi Mat. Nauk* **37** no. 4 (1982), 96. (In Russian.) (Page(s): 162, 167.)

[165] E.E. Landis, Light elements in space–time, passing through a hypersurface in configuration space, *Tr. Sem. Vektor. i Tenzor. Anal.* **22** (1985), 6–68. (In Russian.) (English translation: *Selecta Math. Sovietica* **8** no. 4 (1989), 341–349.) (Page(s): 162, 166, 167.)

[166] V.I. Arnold, Singularities of Legendre varieties, of evolvents and of fronts at an obstacle, *Ergodic Th. & Dynam. Syst.* **2** (1982), 301–309. (Page(s): 163, 170, 193, 199.)

[167] S.M. Voronin, Analytical classification of pairs of involutions and its applications, *Funct. Anal. Appl.* **16** (1982), 94–100. (Page(s): 167, 174.)

[168] S.M. Voronin, Analytical classification of germs of conformal mappings $(\mathbf{C}, 0) \to$ $(\mathbf{C}, 0)$ with identical linear parts, *Funct. Anal. Appl.* **15** (1981), 1–13. (Page(s): 174.)

[169] J. Ecalle, Théorie iterative: introduction à la théorie des invariants holomorphes, *J. Math. Pure Appl.* **54** (1975), 183–258. (Page(s): 174.)

[170] V.I. Arnold, Singularities in variational calculus, *Itogi Nauk i Tekhn. VINITI Sovr. Probl. Mat.* **22** (1983), 3–55. (In Russian.) (English translation: *J. Soviet Math.* **27** (1984), 2679–2713.) (Page(s): 170, 174, 189, 199.)

[171] A.G. Gasparjan, Applications of the higher dimensional matrices to the study of polynomials, *Dokl. Armenian Akad. Nauk* **70** no. 3 (1980), 133–142. (In Russian.) (Page(s): 180.)

[172] O.P. Shcherbak, Singularities of frontal mappings of submanifolds of projective space, *Uspekhi Mat. Nauk* **37** no. 4 (1982), 95–96. (In Russian.) (Page(s): 184–186.)

[173] M.E. Kazarjan, Singularities of the boundary of the space of fundamental systems, flattening of projective curves and Schubert cells, *Itogi Nauk i Tekhn. VINITI Sovr. Probl. Mat. Nov. Dost.* **33** (1988), 215–234. (English translation to appear in: J. Soviet Math.) (Page(s): 184.)

[174] M.E. Kazarjan, Bifurcations of flattenings and Schubert cells, In: Theory of Singularities and some Applications, Amer. Math. Soc., 1990. (Page(s): 184.)

[175] V.I. Arnold, Remarks on Poisson structures in the plane and other powers of the volume form, *Tr. Sem. I.G. Petrovskii* **12** (1987), 1–14. (In Russian.) (English translation to appear in: J. Soviet Math.) (Page(s): 191.)

[176] V.P. Kostov, Versal deformations of differential forms of degree α on the line, *Funct. Anal. Appl.* **18** no. 4 (1984), (Page(s): 191.)

[177] S.K. Lando, Normal forms for the degrees of a volume form, *Funct. Anal. Appl.* **19** (1985), 146–148. (Page(s): 191.)

[178] A.N. Varchenko, Local classification of volume forms in the presence of a hypersurface, *Funct. Anal. Appl.* **19** (1985), 269–276. (Page(s): 191.)

[179] V.P. Kostov, Versal deformations of differential forms of real degrees on the real line, *Math. USSR Izv.* (1990), (Page(s): 191.)

[180] J.-P. Francoise, Modèle locale simultané d'une fonction et d'une forme de volume, *Astérisque* **59/60** (1978), 119–130. (Page(s): 191.)

[181] V.I. Matov, Topological classification of germs of maximum functions and minimax functions of generic families of functions, *Russian Math. Surveys* **37** no. 4 (1982), 127–128. (Page(s): 199.)

[182] V.I. Arnold, Surfaces, defined by hyperbolic equations, *Math. Notes* **44** (1988), 489–497. (Page(s): 225, 229–231.)

[183] V.I. Arnold, On the interior scattering of waves, defined by hyperbolic variational principles, *J. Geom. and Phys.* **5** no. 4 (1988), (Page(s): 225, 229.)

[184] B.A. Khesin, *Singularities of light surfaces and hyperbolicity boundaries of systems of differential equations*, In: Theory of Singularities and some Applications, Amer. Math. Soc., 1990. (Page(s): 225.)

[185] A.D. Weinstein, B.Z. Shapiro, Singularities of the boundary of the domain of hyperbolicity, *Itogi Nauk i Tekhn. VINITI Sovr. Probl. Mat. Nov. Dost.* **33** (1988), 193–214. (English translation to apear in: J. Soviet Math.) (Page(s): 225.)

[186] F. John, Algebraic conditions for hyperbolicity of systems of partial differential equations, *Comm. Pure Appl. Math.* **31** (1978), 89–106; 787–793. (Page(s): 225.)

Index